Investigative Journalism, Environmental Problems and Modernisation in China

Palgrave Studies in Media and Environmental Communication

Series Editors: **Anders Hansen**, University of Leicester, UK and **Stephen Depoe**, University of Cincinnati, USA.

Advisory Board: **Stuart Allan**, Cardiff University, UK, **Alison Anderson**, Plymouth University, UK, **Anabela Carvalho**, Universidade do Minho, Portugal, **Robert Cox**, The University of North Carolina at Chapel Hill, USA, **Geoffrey Craig**, University of Kent, UK, **Julie Doyle**, University of Brighton, UK, **Shiv Ganesh**, Massey University, New Zealand, **Libby Lester**, University of Tasmania, Australia, **Laura Lindenfeld**, University of Maine, USA, **Pieter Maeseele**, University of Antwerp, Belgium, **Chris Russill**, Carleton University, Canada and **Joe Smith**, The Open University, UK

Global media and communication processes are central to how we know about and make sense of our environment and to the ways in which environmental concerns are generated, elaborated and contested. They are also core to the way information flows are managed and manipulated in the interest of political, social, cultural and economic power. While mediation and communication have been central to policymaking and to public and political concern with the environment since its emergence as an issue, it is particularly the most recent decades that have seen a maturing and embedding of what has broadly become known as environmental communication.

This series builds on these developments by examining the key roles of media and communication processes in relation to global as well as national/local environmental issues, crises and disasters. Characteristic of the cross-disciplinary nature of environmental communication, the series showcases a broad range of theories, methods and perspectives for the study of media and communication processes regarding the environment. Common to these is the endeavour to describe, analyse, understand and explain the centrality of media and communication processes to public and political action on the environment.

Titles include:

Alison G. Anderson
MEDIA, ENVIRONMENT AND THE NETWORK SOCIETY

Stephen Depoe and Jennifer Peeples (*editors*)
VOICE AND ENVIRONMENTAL COMMUNICATION

Jingrong Tong
INVESTIGATIVE JOURNALISM, ENVIRONMENTAL PROBLEMS AND MODERNISATION IN CHINA

Investigative Journalism, Environmental Problems and Modernisation in China

Jingrong Tong
University of Leicester, UK

© Jingrong Tong 2015

Softcover reprint of the hardcover 1st edition 2015 978-1-137-40666-8

First published 2015 by
PALGRAVE MACMILLAN

Palgrave Macmillan in the UK is an imprint of Macmillan Publishers Limited, registered in England, company number 785998, of Houndmills, Basingstoke, Hampshire RG21 6XS.

Palgrave Macmillan in the US is a division of St Martin's Press LLC, 175 Fifth Avenue, New York, NY 10010.

Palgrave Macmillan is the global academic imprint of the above companies and has companies and representatives throughout the world.

Palgrave® and Macmillan® are registered trademarks in the United States, the United Kingdom, Europe and other countries.

ISBN 978-1-349-48802-5   ISBN 978-1-137-40667-5 (eBook)
DOI 10.1057/9781137406675

This book is printed on paper suitable for recycling and made from fully managed and sustained forest sources. Logging, pulping and manufacturing processes are expected to conform to the environmental regulations of the country of origin.

A catalogue record for this book is available from the British Library.

Library of Congress Cataloging-in-Publication Data

Tong, Jingrong.
    Investigative journalism, environmental problems and modernisation in China / Jingrong Tong.
        pages cm
        1. Journalism – Political aspects – China. 2. Investigative reporting – China. 3. Pollution – China. 4. China – Economic conditions – 2000– I. Title.
PN5367.P6T659 2015
070.4'30951—dc23   2015001723

To my parents and my hometown

Contents

List of Figures

List of Tables

Preface

Nostalgia and ecological rift

The writing of this book has its origins in my nostalgia for the past and for my native land. As one of the generation born in the 1970s, I have witnessed and experienced two radical changes in Chinese society. On the one hand, I have seen with my own eyes the dramatic transformation of material life that has been happening since the 1980s. I still have childhood memories of a poverty-stricken China. This memory of poverty that still remains fresh in my mind is a sharp contrast to the wealthy and prosperous lives that most of my friends and relatives from my hometown now enjoy. I have witnessed how modernisation has completely turned our lives upside down: people no longer live in wooden houses built in the Ming and Qing dynasties but now live in skyscrapers; people no longer ride clinking bicycles but now drive cars; and people no longer wear black and grey uniforms but stylish fashions. Modernisation is part of the daily life of Chinese people rather than merely an economic policy promoted by the government.

On the other hand, I have also witnessed how modernisation has torn apart and destroyed the environment and natural beauty of my hometown. I grew up in a small farming and fishing water town (*shuixiang*) in Jiangnan (the geographical area to the south of the Yangtze River). When I left home for university at the end of the 1990s, my hometown was still a beautiful place with rich local resources of fruits and vegetables. Twenty years later, however, when I returned to visit my family and friends, I found that the town was no longer the beautiful Jiangnan water town of my memory. Agricultural fields and citrus trees (once a symbol of my hometown) have disappeared; rivers have dried out and been polluted; and almost all period houses built hundreds of years ago have been demolished. Industrialisation – particularly the long-term development of the chemical industry – and urbanisation have taken away my hometown. Modernisation has turned the people who miss their natural, native land into a kind of diaspora. My most fascinating memory of my homeland is walking through endless fields reaching to the horizon, smelling the fragrance of citrus or osmanthus flowers in the evening, or sitting on the bridge in the sunset watching people swimming or fishermen using cormorants to catch fish in the river. However, all of these things that made people feel peaceful and calm in their hearts in the past have already gone with the wind. Concrete buildings have been constructed on the fields. Citrus and osmanthus trees alongside the Xijiang River have been chopped down, replaced by multi-storied buildings or man-made parks that look like a ridiculous parody of nature. Under the five-arch bridge (*wudong qiao*)[1] is no longer a river of clear water but some stagnant, smelly black fluid. The town has already become a forest of steel, brick and cement. I am overwhelmed by the feeling of strangeness and sadness every time I return to it or hear news about it. Sad news about relatives and friends who have been diagnosed with cancer has also come to me from time to time. Cancer seems to have become an epidemic. Of course, the cause of these cancers may or may not be the environment, although people certainly blame the environment.

Modernisation and the preservation of the natural environment have turned out to be incompatible with and contradictory to each other in my hometown. While many of the townsfolk have accepted modernisation as part of their lives, it is obvious to me that the old peaceful agricultural life is over and a violent industrial era has replaced it. Improvement in people's material life has been accompanied by the deterioration of the living environment and shocks caused by that deterioration. I wondered whether this contradiction was unique to my hometown.

Back in 2006, I joined two journalists in investigating a mine disaster in Shanxi Province. The prosperity of the local coal and mining industry was obvious, and I saw an endless stream of trucks carrying coal rushing noisily along the roads and many small cinders floating in the grey air, visible even to the naked eye. I felt suffocated and was unable to open my eyes as they were dry and stinging. In the same year, during a reporting tour of the city of Dongguan in the wealthy province of Guangdong, a noted centre of manufacturing, I talked to "stone-lung" (silicosis) patients and understood their suffering. They were dying helplessly from a disease caused by exposure without proper protection to dust in factories or mines. On other occasions, I met with cancer victims who were unable to obtain justice or leave the environment that they believed had caused their misfortune. They started to become aware of the problems of the environment when their health deteriorated, although initially they may have been happy to receive some improvement in their incomes from the launching of enterprises in their villages. Their descriptions of their miserable lives and suffering led me to understand that environmental problems were not limited to my hometown. Instead they are prevalent all over China, like a huge tumour engulfing the country.

Judging from this personal observation, a clear rift has opened up between modernisation and the health of the environment, with irreconcilable conflicts between the two. Modernisation appears to have become a powerful force that is destroying the environment, and it seems as if other forces are unable to oppose and stop it. This is the ecological rift between "humanity and nature" as defined by Foster and his collaborators (Foster 1999b; Foster, Clark et al. 2010: 7). By "the ecological rift" they mean the rift between modern industrial capitalism and the environment, the notion that the destructive force of capitalism inevitably leads to environmental deterioration. For them, the rift exists because industrial capitalism breaks down "the metabolic relation between human beings and nature" and is, in the long term, unsustainable (7–8). This is an extremely pessimistic view of the future of humanity, as it suggests that the environment will be unavoidably damaged as long as the capitalism that underpins modernisation continues. In China, the situation may be even worse, given its practice of state capitalism. Capitalism in China progresses in an unbridled manner, with the whole-hearted support of the state, in the name of modernisation. If they are right, the environment in China can never again be what it was.

Note

1. A local ancient bridge in my hometown that was built in the Qing Dynasty, about five hundred years ago.

Acknowledgements

I am indebted to many people who have contributed in many ways to this book. I am grateful to friends in the media and NGOs for sharing their thoughts with me, in particular Xiaodong Bao, Haidong Cao, Ke Cao, Chenxing Chen, Lewei Chen, Xuebing Chen, Fei Deng, Zhixing Deng, Sanwen Fang, Tiantian Fang, Hongping Feng, Jie Feng, Jianfeng Fu, Min Guo, Haining He, Xin He, Lu Hua, Yunyong Jia, Guangzhou Jian, Gong Jing, Bo Li, Hujun Li, Jun Li, Qiyan Li, Tingzhen Li, Xinzu Li, Zishan Li, Tianhong Lin, Xiaoqiong Lin, Jianfeng Liu, Ying Liu, Zhi Long, Bin Lu, Hui Lu, Changping Luo, Minghe Lv, Bo Meng, Xianghong Nan, Jinsong Pan, Jiaoming Pang, Hongliang Ouyang, Qing Qin, Renwei Tan, Yaoguo Tang, Peng Tian, Shaochen Shi, Shifeike, Chengbo Wang, Keqin Wang, Jilu Wang, Lei Wang, Peng Wang, Yuchen Wang, Xing Wang, Banyong Wei, Huabinh Wei, Wuhui Wei, Chuanzhen Wu, Chuanfang Wu, Ke Wu, Chuanmin Yang, Haipeng Yang, Xiaohong Yang, Chen Yu, Xiaobing Yuan, Hongwei Yin, Luhuai Xiao, Chunxin Xu, Nan Xu, Zhihui Xu, Dongfeng Zhang, Ke Zhang, Wei Zhou and Hongjun Zhu. In addition, I would like to thank *Southern Weekend* for permitting and sponsoring me to attend its "Green Media Development Program" annual meeting in 2011. I benefited from the discussions in the meeting. I thank Ji Lin and Jingwei Tong for transcribing interviews. I thank Sue Sparks for patiently editing the first draft.

I am grateful to Anders Hansen for inviting me to contribute to this series and encouraging and supporting me throughout the process. Thanks to the anonymous reviewer who provided valuable comments. Thanks to Felicity Plester, Chris Penfold, and Sneha Kamat Bhavnani at Palgrave Macmillan for supporting the project throughout.

The work was funded by the British Academy (Ref: RV100029) and by the College of Social Science and the Department of Media and Communication at the University of Leicester (Research Development Fund). The writing of the book was supported by a period of academic study leave granted by the University of Leicester in the autumn of 2013.

I would like to thank Taylor & Francis for permission to use Figure 4 and Tables 1–3 in the article "Environmental risks in newspaper coverage: a framing analysis of investigative reports on environmental problems in 10 Chinese newspapers", in *Environmental Communication*, 8(3), 345–367. Another figure was generated in the World Bank databank.

I thank Landong for his support and precious comments and am especially grateful for the company of Daniel while writing the book.

Introduction

Modernisation, hegemony and discourse of risk

China faces enormous environmental problems after having gone down the road of modernisation for decades. Modernisation has become a pervasive hegemonic discourse through the process of economic reform in this country. It is everywhere, penetrating into every pore of Chinese society. Naturally, it has become an integral part of China and an everyday term appearing in officials' speeches, the news media, daily conversations and school textbooks. Everyone seems to have accepted it, has praised it and is living by it. Modernisation has become taken for granted among the Chinese people: modernisation is what we should strive for. If the goal of modernisation were to be taken away, we would feel a loss and see our lives as hollow. In Gramscian terms (Gramsci 1971), the existence of hegemony depends on whether or not the ruled can naturally accept the rule of the ruler. This is the origin of power. At this point, modernisation has become a hegemonic power, convincing all Chinese people to accept its rule. It has also become a grand narrative that is accepted as an ultimate goal, one that cannot be questioned, doubted or opposed, and with which no other discourse can compete. This is a driving force, a guideline and a supreme ideology for the development and future of China.

For a long time I have been asking myself the question: while modernisation is a pervasive hegemony which has put the nation's environment at risk, are there any counter-hegemonic powers that can be deployed on the side of the environment and can resist the power of modernisation? I began considering the theories of discourse that might be useful for answering this question. I started thinking about the role of discourse in opposing the hegemony of modernisation, that is, whether

1

the discourse of environmental problems can defy the hegemonic power of modernisation.

Fairclough, Fowler, van Dijk and Foucault share the view that discourse is closely related to power. Though agreeing on the significance of discourse to power relations, they take different approaches to understanding and examining discourse. Foucault suggests that the importance of discourse lies in its exercising of "bottom-up" power/social control through offering a powerful means of discipline (Hannigan 2006; Foucault 1979). It is the "microphysics of power" and "capillary power" that is exercised through and defined by discourse (Foucault 1979; Morris 2012). Foucault believes that power originates from the self-discipline enabled by discourse (Foucault 1979). For him, the most powerful and effective power is the power generated by self-discipline, as exemplified in his perfect power mechanism of the Panopticon. When certain discourses are historically formed in a social context, standards and rules appear, becoming porous, bottom-up and subtle powers regulating people's behaviours (Foucault 1979, 1972, 1970). While placing an emphasis on the formation of discourse, Foucault nevertheless is criticised for ignoring the implications an analysis of actual texts could offer (Foucault 1972; Cao 2001; Brand, Thomas et al. 2005). Instead, Foucault sees discourse as a system of statements and that it is about "language and practice" (Hall 1992), while overlooking the discourse generated in and by text (Brand, Thomas et al. 2005).

In contrast, Fairclough (1989, 1992, 1995), Fowler (1991) and van Dijk (1988) promote an approach that looks into analysing the semantic and linguistic features of texts, taking a dialectic relationship between discourse and existing social reality into account (Brand, Thomas et al. 2005). Fairclough, Fowler and van Dijk thus fill in the gap left by Foucault. For them, the discourse emerging in texts is important and has the potential for transforming society, though its meaning needs to be understood and interpreted by situating it within a specific social context and by considering its interaction with social practices (Fairclough 1992, 1995, 1989; Fowler 1991; Van Dijk 1988). For them, texts and language that need to be considered within external social contexts can reveal important meanings for our understanding of discourse and its relationship to the societies within which it resides. Discourse arising from and in texts is shaped by power and is itself power.

Both approaches are useful for examining and understanding environmental discourses in news media coverage. The four scholars' views on the power of discourse are on two levels of power. Their views are not contradictory but can be integrated to contribute to our understanding of power: we can start from the level of text to the level of society to understand the power of discourse. Although we cannot take for granted the power of text, as readers may resist the dominant meanings of text, discourse that rises from text contributes to the shaping of social discourse on certain issues and plays its part in social change. Discourse not only shapes power relations in society but also mirrors existing power relations, as discourse itself is a site of negotiating power. Discourse has power when the knowledge provided by it changes people's views of certain things and thereafter their behaviour.

In other words, the gist of environmental discourses can be grasped by analysing the linguistic features of media texts and understanding their meanings within social contexts. Both discourse analysis and framing analysis provide two important analytical vehicles for achieving such a purpose (Eder 1996). Meanwhile, Foucault's discourse analysis approach endows us with a means of understanding the formation and power of such environmental discourses more systematically. In other words, an understanding of environmental discourses can be gained first from analysing the texts of these reports and then through examining the meaning of the discourses in the wider social context. The meanings of the discourses are generated in the course of discussing the interplay among them and the social reality outside them. In so doing, it is possible to obtain an insight into the formation of environmental discourses in the coverage of the news media and in society and the role these discourses play in China's modernisation.

An important implication of the four scholars' views on discourse and power is that it offers a way to answer the question of whether there is a discourse on environmental problems (as well as of the relationship between modernisation and the environment) that can act as a counter-hegemonic power resisting the hegemony of modernisation. That is to say, whether and how the hegemonic myth of modernisation can be broken down depends on whether there is a discourse forged within Chinese society describing the dangers and risks brought about by modernisation. In other words, what kind of discourse concerning the relationship between modernisation and the environment, and, especially, the relationship between human development and the environment, is being shaped in Chinese society and is this discourse strong enough to compete against the grand narrative of modernisation?

This kind of discourse is a discourse of risk. From the perspectives of social scientists, such as Beck (1992, 1996; Beck, Giddens et al. 1994) and Giddens (1991, 1990), modern capitalist industrialisation has led to or will result in the emergence of the risk society. For them, risk is an outcome of capitalism (Waisbord 2011). In a risk society, the appearance and severity of risk is beyond the capability of governments and scientists to handle and no one can escape from risk in a risk society. Technologies are also unable to solve the problem of risks that are created by the use of technologies in industrial societies. Beck acknowledges the centrality of the news media in keeping alive both public understanding and public worries about risk (Lester 2010). He regards the shift from an industrial to a risk society as manifested in the embedding of risk discourse in public debates. News media are the conventional sites where public debates take place and the main vehicles that make the general public aware of the existence of risk. As a result of this, news media play an important part in the shaping of risk discourse and the rise of the risk society.

Risk discourse means the interpretation of risk, such as the nature of the risk (what is the risk?), the cause of risk (what causes risk), the generators of risk (who should be responsible for the risk) and the victims of risk and so on (Hansen 2010). In this sense risk is seen as "a product of knowledge", as different understandings may portray different versions of risk (Waisbord 2011). "Risk should be seen as a joint product of *knowledge* about the future and *consent* about the most desired prospects" (Douglas and Wildavsky 1982: 5, original italics). Waisbord regards this perspective as the social constructionist perspective. Waisbord's article (Waisbord 2011) discusses the distinction between Beck's and Giddens's risk society perspective and the social constructionist perspective (such as that of Douglas and Wildavsky 1982) towards understanding risk. Nevertheless, I argue that there can be a convergence between the risk society and the social constructionist perspectives of understanding risk. The importance of risk discourse in the transformation from an industrial to a risk society acknowledges the significance of the interpretation of risk, although there may be a variety of interpretations of risk that reflect ideological conflicts among various interest groups. This does not mean risk does not exist but instead is an artefact of discourse. The existence of risk is an objective phenomenon, whereas the discourse of risk is subject to the interpretations of risk. One important interpretation of risk is the one offered by the news media. It is the discourse of risk shaped by the media interpretation of environmental problems in Chinese society that is the focus of this book.

The media landscape and investigative journalism in China

Investigative Journalism, Environmental Problems and Modernisation in China examines how the news media in general, and investigative journalism in particular, interpret environmental problems and how those interpretations contribute to the shaping of a discourse of risk that can compete against the omnipresent and hegemonic discourse of modernisation in Chinese society. News media are not free in China. Political control over news media remains unchanged, although media commercialisation and marketisation has taken place under political administrative commands since the 1980s (Zhao 1998). Facilitated by media marketisation and the rise of new media – especially the Internet, Web 2.0 tools and mobile electronic devices – Chinese news media have achieved certain levels of journalistic autonomy and have manoeuvred to create a certain amount of space in which to function. The proliferation of commercial news media in the 1990s and early 21st century offered fertile soil for the development of journalism, especially investigative journalism. Nevertheless, in spite of having brought a certain level of autonomy to news media, especially in the early stages of media marketisation, the introduction of the market has turned out to be a constraining force as a result of the practices of state capitalism in the media sector (Lee, He et al. 2006, 2007). The drive for maximal market profits by news media leads them to giving up their dream for media freedom in exchange for the support of the Party-state and its allies in the business sector. Entering the second decade of the 21st century, when the market has become increasingly monopolised by giant media conglomerates, media control has been tightened instead of loosened, especially in the later reign of Hu-Wen and under the leadership of Xi.

The Hu-Wen leadership showed a benign and friendly attitude towards news media at the start of their reign, especially during the outbreak of the epidemic disease SARS in 2003. However, later on, the leadership turned to squeezing the space available for news media to practice independent journalism. This has been manifested in continual crackdowns on any news media that run investigative reports that bravely annoy the political authorities and their allies in the economic sector. Prominent examples include the removal of editors-in-chief at *Southern Metropolitan Daily (nanfang dushibao)*, *Southern Weekend (nanfang zhoumo)*, *Beijing News (xinjingbao)*, *Xiaoxiang Morning (xiaoxiang chenbao)* from 2003 onwards, and the exile or banning of media commentators and activists such as Chang Ping (Zhang Ping), Xiao Shu, Bei Feng (Wen Yunchao), Mo Zhixu and Chen Guangchen. A regulation forbidding cross-regional investigative reporting was issued by the central government in 2004.

At the time of writing this book, Xi Jinping acceded to the presidency. Since the start of his leadership in late 2012, a serious of events such as online "purges" and crackdowns on mainstream news media as well as his speech on propaganda and thought work in China in the

National Propaganda and Thought Work Conference in August 2013[1] have conveyed a signal that Xi's media policy intends to constrict rather than relax media control. The era of Xi has therefore started with severe restrictions on news media and freedom of expression in 2013 and 2014, as exemplified in the arrests of Weibo bloggers such as Xuan Manzi and Wang Gongquan, who have a huge number of followers on the Internet, the warnings given to influential commentators and journalists such as Li Chengpeng and Luo Changping, as well as the crackdowns on news media like *Xinkuai Daily* (*xinkuaibao*).

The situation is further worsened by news organisations' desire for maximal profits, which pushes them to collaborate with governments for guaranteed profits and beneficial policies and with economic institutions for mutual commercial benefits. Seeking collaboration with the political authorities is an effective and safe strategy for the traditional Chinese news media to continue to be profitable, especially when they are suffering huge financial losses in the market in the second decade of the new century as a result of the challenges presented by new media. The financial crisis started in 2012 and became prominent in 2013 (Zhao 2014). For example, in March 2013 *China News Media and Publishing* newspaper (*zhongguo xinwen chuban bao*) published a news story declaring that the advertising income of new media overtook that of the traditional news media in Shanghai for the first time.[2] Newspapers such as *News Evening* (*xinwen wanbao*) and *Tiantian Xinbao* (*tiantian xinbao*) in Shanghai closed in 2013 and 2014. Traditional news media are facing a harsh situation and need to work harder for their survival in the market.

The financial difficulties of Chinese news media have created an expectation of governments to give them financial support. Back in 2009, for example, with no consideration of the impact on the independence of the Chinese press at all, Chinese scholars like Cao Peng even appealed to governments for support (in the form of providing financial subsidies, subscribing, offering beneficial taxation policies and lifting bans on launching newspapers and so on) to save the press from decline (Cao 2009). In the following years, the revival of financial subsidies offered by political authorities to support local media groups has proved the tendency for Chinese news media – especially Party organs – to turn to the Party for financial aid. A prominent example of this was that the Shanghai government promised in 2013 to annually subsidise *Jiefang Daily* (*jiefang ribao*) and *Wenhui Daily* (*wenhui bao*) (both are Party organs) RMB 100 million (Fan 2013; Zhou 2013).

Under such circumstances, therefore, talking about whether the Chinese news media can construct an anti-modernisation discourse sounds like a ridiculous joke, given that modernisation has been China's top priority since the establishment of the People's Republic. However, the reality is much more complex and contradictory than this might suggest. Neither the news media nor discourse in China is a monolithic bloc. The media–government relationship is complicated, marked by de-centralisation and by changes over time along with shifts in social dynamics (Tong 2010). Geographical variations in journalism cultures have been recognised (Tong 2013). Multiple voices are competing in present-day China, one of which is about the relationship between modernisation and environmental problems. One of the main vehicles for different and even critical voices is investigative journalism.

Investigative journalism in China often involves a large investment of time and money. In general, the aim of this type of journalism is to expose events that the privileged would like to hide from the eyes of the public and which the general public are unaware of otherwise. It also strives to discover the truth of how things come to be the way they are and to tell its readers what has happened and why, after having conducted extensive investigation (Tong and Sparks 2009; Tong 2011). Rather surprisingly, the rise of investigative journalism in the 1990s was due in large part to the need of the central government to conserve inner Party purity and to consolidate the legitimacy of its rule (Zhao 2000; Svensson, Saether et al. 2014). Despite this governmental initiative, media market incentives and occupational requirements subsequently combined to drive investigative journalism to become institutionalised within news organisations such as *Southern Weekend*, *Southern Metropolitan Daily* and *Beijing News* and to be occasionally practised in other news outlets such as *Yunnan Information* (*yunnan xinxibao*) (Tong and Sparks 2009). Investigative journalism has gone on to be practised on a national scale since the mid-1990s, in spite of crackdowns from political authorities on some news media. While the flames of this type of journalism have died down in some news organisations, such as *Dahe Daily* (*dahebao*), its spark continues within other news organisations (Tong 2013). Investigative journalism is a genre of journalism having social significance and the potential to advance social change. Particularly in China, where the news media are under tight control, this type of journalism provides possibilities for diverse voices to be heard and for constraints on the abuse of power by governments.

The book describes investigative journalism on the subject of environmental problems as "environmental investigative journalism". Developed from the definition Wyss has given to environmental journalism (Wyss 2008), environmental investigative journalism in this book refers to a genre of investigative journalism that investigates and explores environmental problems, risks and hazards. This genre of investigative journalism requires normal investigative journalistic skills as well as special abilities to write about abstract and obscure scientific knowledge in simple and easy-to-understand language. Therefore, environmental investigative reports often involve an interpretative and informative content. In addition, witness accounts of scenes and situations are often necessary to such reports. Environmental investigative reports involve more extensive investigation and are more expensive and lengthy than daily environmental reports. Like investigative reports on other topics, environmental investigative reports usually do not need to meet the daily deadline of submission and go beyond the normal "green" beat. Therefore environmental investigative journalism enjoys more flexibility in terms of topics, research, submission deadlines and length than daily environmental journalism. In addition, environmental investigative journalism usually involves a lot of scientific knowledge and on-the-scene description which investigative reports on other topics might not include.

The attention given by investigative journalists to environmental problems has increased along with the development of investigative journalism itself and the growing prominence of environmental problems. This was first seen at the end of the 1990s in news outlets such as CCTV's *Focus* (*jiaodianfangtan*) and then shifted to appear in the coverage of the print media. For example, while acknowledging the important part *Focus* played in the war against deforestation in the late 1990s, one cannot ignore the significant role played by newspapers and magazines with a long-established investigative journalism history, such as *Southern Weekend, Southern Metropolitan Daily, First Financial and Economic News* (*diyi caijing*), *Caijing* magazine and *New Century* (*xinshiji*) magazine in the anti-dam-construction agenda setting and development in the new century.

An overview of investigative reports on environmental problems reveals that investigative journalism has given continual and enduring attention to environmental problems and relevant agendas since the end of the 1990s, although diverse agendas have appeared at different periods during this time. The end of the 1990s and the early 21st century has seen investigative journalism develop into a mature genre of journalism that treasures objective reporting. Compared with those on other topics such as social and political issues, investigative reports on environmental problems are unique. They are unique for three reasons. First, although also involving the revelation of scandals and wrongdoing by individuals and institutions, environmental investigative reports on some occasions stress the interpretation of environmental problems in terms of causes, consequences and current conditions as well as highlighting the presentation of different views surrounding particular issues. Therefore, environmental investigative journalism is more interpretative in this sense. Second, environmental topics enjoy much more reporting autonomy than other topics. Having more autonomy means not only that the political authorities give more permission for this type of investigative reporting but also that the evidence of environmental problems is out there and can be witnessed and observed by journalists who do not have to rely on official or expert news sources for providing them with information. Third, this can be seen as the only type of investigative reporting that is welcomed and sponsored by commercial enterprises, although it criticises some enterprises for destroying the environment. No other type of investigative reports have any of these traits.[3]

From 2011–2013, the author interviewed 42 nationally well-known environmental investigative journalists from influential news outlets. "Environmental investigative journalists" refers to investigative journalists who have constant interests in environmental problems and issues and have already produced important investigative reports on this topic over recent years. My interviews with them show an interesting commonality: most of these journalists[4] interviewed were born in the 1970s and 1980s. Like me, they have experienced and witnessed the process of modernisation during which the environment has been gradually deteriorating. A shared understanding among us is the incompatibility between modernisation and the environment. Given the advocacy nature of investigative journalism and the personal experiences and cognition of the relationship between modernisation and the environment, investigative journalism on this topic is a site that is very likely to produce a counter-hegemonic discourse. Therefore, on the one hand I am trying to distance myself from my life experiences and those of investigative journalists and objectively scrutinise investigative journalism (trying to avoid my personal biases) and on the other I hope the examination of this type of journalism can offer some interesting answers to the question that has been on my mind for quite a long time. The questions I intend to discuss in this book are: What kind of risk discourse has been constructed by investigative reports on China's environment problems? Whether and to what extent can this discourse of risk facilitate the forging of a counter-hegemonic force against the hegemony of modernisation?

Answering these questions involves two levels of understanding of the discourse of risk. The first level refers to understanding the interpretations of environmental problems as well as the nature of views on the modernisation-and-environment relationship embodied in the risk discourse. Specific interpretations of environmental problems, defining risk and reflecting perspectives on the modernisation-and-environment relationship, function as the basis from which risk discourse is constructed. It is not unusual to see the coexistence of various types of views even in the same society. But during a specific period of time, one type of view may possess a dominant position and thus override other types of views. In China, to take an example, a long-established environmentalism, shaped by history, that treasures the harmony between humanity and nature (*tianren heyi*), is out of tune with the currently prevailing view, embodied in modernisation, that favours economic development over environmental protection. But the rise of alternative environmentalisms, of course, has the potential to present a challenge to the dominant view favoured by the Party-state if it is strong enough. Therefore, it is crucial to find out what kind of views of the relationship between the environment and human activity is embedded in the risk discourse constructed by investigative journalism. Does the risk discourse take the side of modernisation or oppose it on behalf of the cause of environmental protection? In order to answer these questions, the study has examined influential environmental investigative reports published since the end of the 1990s (see Chapter 2). In addition, the study has also adopted framing and discourse analysis to analyse in detail 258 investigative reports covered in ten newspapers from 2008–2011 (see Chapter 3).

On the second level, the understanding of the risk discourse is located within the context of news production. The practices and norms of investigative journalism and the ecology within which it operates require attention in the analysis. Journalism is a "principal convenor" of discourses and debates (Cottle 2006: 3). While shifts in social contexts and organisational cultures impact on investigative journalism and its practices and norms, the reports produced by investigative journalism are also influenced by its practices and norms and any changes in them. Apart from that, journalistic work also takes place in the context of the interaction with power dynamics in society. This interaction is a process through which powers struggle for the right to define issues and reality and within which different, or even opposite, viewpoints compete for publicity (Lester 2010). Without a doubt, this process is in part responsible for the nature of risk discourse. It is thus important to take into account the role of journalistic work in shaping the risk discourse, helping to explain why the risk discourse appears in the way it does.

Social dynamics

This exploration of the two levels of understanding of risk discourse has to take place within the context of Chinese society (especially social dynamics and the relationship between the state and society). At present, Chinese society is full of tensions and dynamics that both domestic and overseas observers are endeavouring to understand. After the Tiananmen Square event in 1989, China has seen no political reform. The CCP has continued its authoritarian rule. Nothing radical and fundamental has happened in the political arena since the 1980s' economic reform. Despite that, a broad spectrum of ideas has emerged in society. Leftists, rightists, neo-leftists and neo-rightists have already had fierce debates on the future of China and the consequences of the economic reforms (for example, see Wang 2006, 2005; Wang and Karl 1998; Xiao 2002, 2003; Li 2002). These debates largely focus on changes in the economic and social spheres. Especially in recent years, China has also seen a surge in thinking with political implications. These include reflections on the Cultural Revolution,[5] the merits and sins of Mao Zedong, the Great Famine (1959–1961) during the Great Leap Forward movement, and even the contributions of the Kuomintang Nationalist Party during the anti-Japanese war. All of these topics were absolutely forbidden in the past. The emergence of discussions of this kind not only signals that Chinese society is not a quiet backwater at all, but also suggests possible changes in politics.

Accompanying these diverse ideas are intensive social conflicts and tensions caused by widening social inequalities, the opaque and incomplete judicial system and procedures, the lack of social justice, soaring house prices, urban demolitions, antagonism among different classes and even among occupations, such as patients and doctors, vendors and urban administrators (*chengguan*),[6] as well as deteriorating ethnic relations, especially between Han and Uighur. A series of bloody and violent events, such as several incidents of patients killing doctors,[7] a number of self-immolations, cases of arson or murder in relationship to land acquisition and urban demolition,[8] and several killings of vendors and of urban administrators[9] demonstrate and offer primary examples of the extremism fostered by these tensions and conflicts. On top of all these social dynamics and tensions is the emergence of social protests, such as environment-related protests, protests over freedom of speech, workers' protests, social injustice–induced protests, protests against land acquisitions and urban demolition, and protests over ethnic issues that strongly demonstrate disagreement or dissatisfaction over certain issues.

Among these protests, ones over environmental issues have been most frequently seen over recent years and tolerated the most by political authorities. This is for two reasons. First, environmental issues are of concern to all classes in society. Second, environmental protests usually target economic enterprises and do not touch politically sensitive issues such as political reform, medical reform, the conflicts between governments and people and so on.

Given the changes that have happened in Chinese society since the 1980s economic reform, scholars have started discussing whether and to what extent civil society has appeared in China, although hope for political reform has been dim since the Tiananmen Square event in 1989. In general, civil society (an important concept in current political thinking in the Western world) refers to an arena (outside the state) in which the people can forge a bottom-up grass-roots force that opposes or constrains the power exercised by the state, which is vital for ensuring the health of democracy (Hodgson and Foley 2003; Kaviraj and Khilnani 2002; Habermas 1989). Civil society can include like-minded private individuals, organisations and a space where these individuals and organisations can voice their needs and organise their activities and where diverse or even contentious interests, thoughts and ideas can be presented and contested. Civil society is counter-hegemonic in some senses, given that civil society provides a space in which dominant ideologies and values are contested (Gramsci 1977).

Scholars have provided evidence for the view that China has seen the emergence and development of civil society, drawn from their observations of the rise of non-government social organisations, the opportunities offered by the Internet, and social movements, in particular environmental movements (Wu 2003; Yang 2003a, 2003b; Howell 2012; Yu and Guo 2012; Simon 2013; Ho and Edmonds 2008). Capitalism is seen as a crucial mechanism that helps to lay the basis of civil society in China (Howell 2012), although capitalism is often the target of civil rights movements. As a result of the introduction of market forces into Chinese society, a series of social dynamics have occurred. Among others, the establishment of civil organisations like NGOs and labour organisations, social groups and associations add liveliness to society. The frequent movement of labour from rural to urban areas and across regions is breaking down traditions and conventional values in localities and thus creating room for new customs and cultures. In addition, Chinese society has also seen active participation from elites like business people, solicitors, journalists and public intellectuals and from ordinary citizens in advancing public issues and in shaping an array of social ideas. The Internet has offered a symbolic and empowerment platform for private individuals and grassroots groups and organisations to nurture their community and skills and to organise and mobilise their actions. Workers' associations such as Foxconn Workers (*fushikanggongren*) and Home of Shenzhen Little Grass Workers (*shenzhenxiaoxiaocaogongyou-jiayuan*) have emerged in Guangdong, the workshop of the world, and have actively used the Internet to attract the world's attention and to protect their interests (Qiu 2009). Activities aimed at protecting workers' rights, such as the recent strike in Guangdong Province in 2014,[10] express workers' voices and the willingness to participate which offer grass-roots and potentially revolutionary resistance against capitalists who are often under the protection of local governments. In addition, the proliferation of environmental movements, among other contentious activities, is one of the main reasons that makes scholars tend to believe civil society is coming into being in China (Yang and Calhoun 2007; Tilt 2009; Lu 2007; Ho 2001). Over the past few years, social protests associ-ated with environmental problems have occurred frequently across the country. According to these scholars, the popularity of domestic and international NGOs, civil movements, workers' movements and heated public debates all indicate the rise of civil society.

However, the concept of civil society in China is different from that in other parts of the world such as Poland, as argued by Béja (Béja 2006). Béja regards the concept as used in the Chinese context does not mean the democratisation of the region nor that China will take the same evolutionary path as that taken by Eastern Europe. For those scholars holding an optimistic view about the development of civil society in China (such as Ho 2001; Spires 2012 and others cited above), the concept of civil society is merely equivalent to social organisations or social move-ments, ignoring other aspects of this concept such as democratisation. The discussion and meaning of the concept of civil society in the context of China has deviated from that in the Western context and history.

In addition, following the Gramscian understanding of civil society (Gramsci 1977; Gramsci's theories have been adopted to examine the state–society relationship in Asian countries such as the studies by Miles and Croucher 2013 and O'Shannassy 2009), one can argue that even if a civil society does arise in China, although it may have counter-hege-monic potential, it could actually help the state to consolidate its legiti-macy and hegemony. Gramsci regards the existence of civil society as constitutive of the state's hegemony, as the state needs to show a stance of tolerating resistance from society. As environmental problems have become an issue of unprecedented importance that cannot be ignored and urgently require the attention of the ruling Party, it is arguably better if the government takes a proactive attitude in responding to anxiety among the public and allows room for the public mood to be expressed rather than trying to completely ignore the problems and attempt to silence people. The central government has been found to have "growing awareness of the need to strengthen civil society" in order to mitigate the pressure caused by social issues such as unemployment and environmental problems (Ho 2001: 902). In this sense, the state's permission for the emergence of social organisations and environmental movements is a means through which the state shows a positive image to the world as well as letting social organisations share the responsibilities and pressures caused by environmental problems. The Party-state's tolerance of the boom in NGOs is believed to "help the party by mitigating social anger and offering health care, education and other services which the party finds it hard to provide".[11] Ho criticises the heavy reliance of China's NGOs on the state for legitimacy and therefore argues that they lack "intimate linkages with citizens and international donors" (Ho 2008: 2). The distinction between the state and civil society in China is unclear and environmental activism in China is deeply embedded in the Party-state and in the semi-authoritarian context (Ho 2008).

The idea of civil society thus seems to have some flaws in China. Nevertheless, no matter what motivation the state has for encouraging the appearance of NGOs and the shaping of civil society, of importance here is the possible change that the Chinese prototype of civil society might be able to bring to society. If we reject the idea of civil society as a term in social science with a fixed meaning and instead see it as a force that is able to generate some real changes in society and in politics, then civil society exists and possesses positive potential in China. This is because any changes would be positive, no matter how small and trivial they might be, given that change of any kind in politics in present-day China, dominated as it is by state capitalism and ruled by authoritarianism, is very difficult to achieve.

The structure of the book

This book examines the role of investigative journalism in the shaping of risk discourse and in representing the relationship between the environment and modernisation against the backdrop of Chinese society. The book is structured in six chapters. Chapter 1 explores the relationship between China's modernisation and environmental problems, revealing the utopian goal of modernisation for "a better life" and the increasing importance of environmental issues in the process of modernisation. It also examines the responses of various parties – namely the state, business and society, with their respective interests – to pressing environmental problems. It shows how two socioculturally rooted logics as to the relationship between human development and nature underlie the environmental situation and the modernisation process. A more detailed look brings to light grass-roots resistance from NGOs and citizen activism against top-down political and corporate power. A wide range of cases taken up by both citizen and NGO campaigns have seminal potential for improving environment protection. Despite that, the capability of society to resist is limited by social reality in China, including the current capitalist production mode, the prevailing consumer culture and lifestyle and the distinctive respective interests of different classes. There is a relatively long history of the mainstream media devoting attention and coverage to the environmental problems emerging in the process of modernisation. The chapter aims to provide a framework for the construction of a risk discourse by investigative journalists. The extent to which the Chinese news media generally – and in particular investigative journalism – mediate the different interests representing various centres of power will be a question raised at the end of this chapter.

Chapter 2 demonstrates that the development of environmental problems into a major topic for investigative journalism over the past 20 years has taken place through a process of interaction between investigative journalism and diverse social interests. The reporting on environmental problems over 20 years has constructed nine agendas, ranging from pollution to artificial land and island construction, classified into three categories: fighting for nature, fighting for human well-being and fighting for both nature and human well-being. The construction of these nine agendas portrays a clear picture of what has happened to the environment, ringing warning bells for the nation. A careful examination of these nine agendas suggests that a contradictory role has been played by investigative journalism in the whole process. On the one hand, by reporting on specific environmental problems, investigative journalism surprisingly functions as an "ideological state apparatus" of the Party-state and, perhaps indirectly, serves the interests of both the state and economic businesses. The exposure of environmental problems surprisingly caters to political and commercial needs. On the other hand, however, investigative reports on particular environmental problems echo and give vent to diversified voices in society, some of which strongly oppose the dominant interests of the state and commercial enterprises. This critical stance of investigative journalism indicates its advocacy role.

Chapter 3 offers an account of the discourse of environmental problems as constructed in the coverage of the Chinese press from 2008–2011. The selected investigative reports are analysed both quantitatively and qualitatively. The focus of the analysis is on deconstructing the way in which environmental problems are presented, for example: what kinds of environmental problems are revealed; the causes and consequences of these problems; who is suffering from them; who should be held responsible; what are the potential solutions and so on. This chapter asks how environmental problems have been interpreted as risks, and whether and to what extent such risks result from China's modernisation and therefore stand in potential opposition to the discourse of development associated with that. These questions are crucial for an understanding of the nature of the press discourse of environmental problems, as well as the potential role of investigative journalism in influencing China's modernisation policies. Overall, two environmental discourses – extinctionism and eco-equalism – have been identified and are discussed.

Chapter 4 looks into the way Chinese investigative journalists do their job in reporting on environmental issues and thus provides reasons to account for the features of the discourses, as discussed in the previous chapter. There are two foci in this chapter. One focus is on the epistemology of investigative journalism and the importance of journalists' (pre-existing) knowledge and cognition of environmental problems in the construction of the discourse of risks. The epistemology of investigative journalism demonstrates how journalists get to know what they know in reporting on environmental problems. The other focus is on examining the tactics adopted by investigative journalists in coping with the actions of the different parties who are aiming to influence their work. Struggles over discourse among different centres of power can be observed in the news production process and will be scrutinised in this chapter. But journalists have demonstrated both tactics and strategies in dealing with struggles of this kind. This chapter further discusses how the mindsets of investigative journalists about environmental problems are transferred to the frames in the discourse in the Chinese press and to what extent the journalists' mindsets match the frames that can be identified in the press coverage itself.

Chapter 5 examines the interaction between offline investigative journalism and online environmental campaigns and its implications for constructing the discourse of risk and resisting the hegemonic discourse of modernisation. It discusses the innovative features of environmental movements in the new media era and how investigative journalism handles the opportunities provided by online environmental movements. In the interaction between the two, investigative reports have become the site where civil discourse and official discourse of environmental problems meet. In this way, investigative reports mediate different viewpoints from society, experts and governments. In addition, investigative journalists' own investigations and new frames and themes added to their reports endorse and verify the civil discourse of environmental problems.

Chapter 6 sums up the role of investigative journalism in the social construction of the discourse of environmental risk and discusses the meaning and limitations of such a role. The discourse of environmental risk constructs an image of a "risk society" and of a rivalry relationship between modernisation and the environment for China, which is seemingly in opposition to the discourse of development with its image of a "harmonious society", and with the "China dream" promoted by the Party-state. This presents a dichotomy of discourses within China's process of modernisation. The counter-hegemonic force of the risk discourse constructed by investigative reports has three aspects. First, by constructing the discourse of risk, investigative reports reveal that the hegemony of modernisation is indeed the hegemony of capital and that the state has been hijacked by ruling interest groups and capital. Second, the link between environmental problems and social injustice shatters the claims modernisation makes about utopian social equality and wealth. And third, the discourse of risk constructed in investigative reports points out the fact that the shocks brought by the revelations of environmental problems indeed reflect people's shock towards the rapid shift from an agricultural to an industrial society. The different values and criteria for evaluating the changes in society and the achievements of the Party-state embodied in the discourse of modernisation and the discourse of risk demonstrates the rift between modernisation and the preservation of the environment. The social tensions, conflicts and dynamics in Chinese society exacerbate the anxiety caused by this rift. The chapter also discusses the limitations imposed by the current media landscape on the counter-hegemonic force of investigative reporting.

1
Modernisation, Environmental Problems and Chinese Society

This chapter maps the relationship between the emergence of environmental problems and China's modernisation. It also discusses the responses of various parties – the state, transnational and national businesses, society – to these issues. It suggests that the importance of the emergence of environmental issues should be understood in the light of their relationship to society, taking into account the potency of social dynamics, especially the interaction between global capitalism and domestic power relations, the development of civil participation, social class differentiation, as well as consumer culture and lifestyles, and of the social logic regarding the relationship between development and the environment.

In its ongoing process of modernisation since 1949, China has transformed itself from an agricultural into an industrial society. Parallelling this domestic process, Western capital has continually sought expansion in the global arena. Developments in transportation and information technologies have provided the material infrastructure for such an expansion. With its enormous natural resources and cheap labour, China offers an ideal opportunity for the expansion and ambitions of global capital. The corollary of increased globalisation has been China becoming pivotal to global production chains, assuming the role of the world's factory.

Capitalism's pursuit of capital accumulation and maximal profits by its very nature places it in opposition to the preservation of the environment. Its impacts on the environment are also exacerbated by the Chinese leadership's ignorance of sustainable development and its belief in Man's capability and right to conquer nature. As a result, the environment is threatened by a proliferation of economic activities encouraged in the modernisation process. Concurrently, Chinese society has experienced a series of dramatic social and political changes, such as the Cultural Revolution and the reforms aimed at opening up the country and accelerating growth, which have given rise to social tensions. As part of that process, environmental problems per se are related to power struggles among a variety of parties with particular interests, especially the state, enterprises and society generally. The news media are a principal site in which these power struggles are negotiated. Chinese media have hitherto played an uncertain role, although many media outlets have given enduring and extensive attention to environmental issues. The genesis of this uncertainty can be ascribed to the unique relationship between the news media and various social actors emerging in the authoritarian context of contemporary China.

The chapter starts with an overview of the features of modernisation and then turns to a discussion of the advent of environmental problems as a consequence of modernisation. After that it looks into the way the state, enterprises and society have responded to the worsening environmental situation. A discussion of their responses reveals the mutual influence between social dynamics and environmental problems and how that influence leads to the ways in which environmental problems are seen at the present time. An overview of the attention paid by the news media to environmental problems and issues follows. The chapter concludes by raising two questions: How do the Chinese news media – and in particular investigative journalists – mediate the respective interests of different parties and opposing viewpoints on environmental issues, and what does such a mediating role mean to China's modernisation?

1.1   Modernisation: a utopian dream of "better lives"

As elsewhere in the world, China's environmental problems are highly associated with its modernisation (here I mainly refer to economic modernisation) and industrialisation which are driven by a longing for a better quality of life. Such a desire for a better life was initiated by the leadership from the outset of the establishment of the People's Republic, and quickly embraced by ordinary people. The goal of modernisation, a sweet dream the Chinese Communist Party (CCP) has woven, has come to be accepted wholeheartedly by the people. Replacing the obvious ideologies promulgated during times of war, (economic) modernisation has become a hegemonic force that has cemented an unwritten agreement with the public and reinforced the leadership of the CCP. The hidden meaning behind modernisation is that to unite around the CCP is the best and only way to live a good life. Modernisation encompasses a utopian ideal that narrows social inequalities and eliminates poverty. At that time China had just emerged from international and civil wars. It urgently needed to put the economy back on its feet. People who had suffered war and poverty over a long period were easily convinced that the national priority should be given to economic growth in order to make the country stronger and improve people's lives. Thus it was not a surprise to see the modernisation wave sweeping across the country from the 1950s. Industrialisation, the renaissance of agriculture and relevant scientific and technological innovations were officially promoted with the aim of bringing the economy forward and achieving national development.

Chinese modernisation can be divided into three phases. The first phase lasted from 1949 to the outbreak of the Cultural Revolution in 1966. As a close political ally of the Soviet Union, China took the same socialist road of modernisation, characterised by a planned economy and heavy-industry–oriented industrialisation. This phase witnessed a shift in the relationship between China and the Soviet Union in which China first relied on the Soviet Union to provide aid and later depended on itself for progress. China started to follow a separate path from the Soviet Union and even wanted to surpass it to become the leader of the socialist countries from the late 1950s. The Great Leap Forward was launched in 1958 for such a purpose but proved to be a failure in terms of both industrial and agricultural development (Rozman 1981). For example, one major purpose of this program was to promote collective mechanised agriculture and increase grain production within a short period of time in order to meet the needs of rapid industrialisation. A tragic result of this program, however, was a nationwide famine due to a sharp decline in crop production. An interesting study by Li and Yang drawing on empirical data collected from 1952 to 1977 has revealed that problems with central planning during the period from 1958 to 1961 were responsible for 61% of the decline in grain output (Li and Yang 2005). The launching of the Cultural Revolution campaign signalled the beginning of the second phase of China's modernisation in which the economy was devastated and modernisation was almost halted because of the political trauma.

Modernisation restarted in the 1980s, entering its third phase, and witnessed a shift from a planned to a market economy. The 1980s was a crucial time in which the need by international capital for global expansion coincided with the concurrent requirement for domestic economic growth. The Fordist mode of production encountered a deep crisis of capitalism in Western domestic markets. Western capitalism urgently needed to transform its production mode and required global markets and resources (Castells 1996; Harvey 1990). Domestically, Deng Xiaoping and his allies launched extensive economic reforms to open up and boost the economy, though leaving political reform behind. The Eastern and Southern regions were the first to join the economic reform program, followed by the Western and Midland regions. While industrialisation and rapid and steady economic growth were the main objectives of the reforms, attracting foreign investment became one of the main means to increase the vitality of industries and boost GDP. Entry into the World Trade Organization (WTO) in 2001 further integrated China into the global economy.

The discussion above reveals one prominent feature of modernisation: for most of the time modernisation has meant economic development and industrialisation for China (Ho 2006). Despite the centrality of economic growth, modernisation is closely linked to or even defined by the dominant ideology in China (Wang and Karl 1998; Schwartz 1965). An authoritarian attitude by the ruling party towards modernisation and its determination to transform the lives of the people underpins modernisation, which not only means that giving its people a better life becomes an objective of the Party, but also suggests that this task has to be fulfilled by the Party itself (Schwartz 1965). More than half a century after Schwartz made that comment, the principle of the CCP's authoritarian rule remains unchanged at the present time and the Party continues to assert that it supports socialism and has the aim of establishing social equality among people. Nevertheless, modernisation, that is, the central planning of social life by the Chinese government, is influenced and even limited by market principles and the logic of capitalism. Because of that, the relationship between the state (especially governments at all levels) and capital has become very intimate (Wang and Karl 1998). When modernisation penetrates into the everyday life of the people, the intimate connection between modernisation and the governing ideology means that modernisation underpins the foundations for the rule of the Party. The decades-long priority for modernisation has made it a hegemonic paradigm that legitimates the authoritarian regime. Core values of the state and later the principles of the market are conveyed to the people in order to be accepted by them. Therefore underlying the need for modernisation is the state's hegemonic interest.

Underlying this is an ambivalent social attitude towards nature, promoted by the top leadership. On the one hand, resources are thought of as finite. Elite individuals and institutions, such as governments, should therefore take responsibility for managing finite resources. This attitude is best embodied in the national policy for controlling the growth in population which embodies fears over resource scarcity. Mao once expressed trust in the power of human resources with his famous saying "stronger with more people" (*renduo liliang da*). However, the population explosion from the 1950s to the 1990s caused the leadership to fear overpopulation and worry about resources and economic growth and resulted in the 1979 launch of the notorious "one child policy" that has been in force in recent decades. The Chinese population increased by 57% from 1949 to 1971 (Jing 2013). Such a fear is, in Dryzek's terms, a kind of survivalist discourse of environmentalism (Dryzek 2005). Nonetheless, this survivalist discourse serves the development discourse of modernisation as, after all, the population is being controlled for the sake of development (Shapiro 2001).

On the other hand, there is an absolute confidence in humans' capacity to conquer nature and handle environmental problems as well as in technology that is seen as supreme. The central leadership followed Marx's and later Mao Zedong's view of nature: seeing nature as an object to be conquered and used for production and progress purposes. Marx's writings are thought to have embodied a Promethean attitude towards nature (Giddens 1981), although, also having paid attention to ecological issues such as the metabolic rift (Foster 1999a; Hannigan 2006: 8–9), Marx's critique of the political economy of capitalism suggests the nature of capitalism is unsustainable and incompatible with nature (Foster 2010). Although Marx criticises capital's exploitation of nature, his Promethean discourse expresses confidence in humans' capability to overcome environmental problems and can be used to justify the national priority of economic growth (Dryzek 2005). The Maoist ideology for conquering nature further justifies and catalyses the exploitation of natural resources in the name of developing the national economy. This is well demonstrated in his famous sayings such as "men can conquer nature like the foolish man can move mountains" (*yugong yishan rending shentian*) and "it is endless fun fighting nature" (*yutiandouqilewuqiong*). During the pre-1980s years, this Promethean belief underpinned the planned economy as well as economic campaigns such as the Great Leap Forward and the Four Pests Movement in 1960s that were initiated by Mao and other political leaders (Shapiro 2001). This belief has remained intact over the past decades.

Indeed, spurred by the overall national priority for economic growth, China's annual GDP has seen a steady increase (see Figure 1.1). From 2010, China has boasted of possessing the world's second-biggest economy.[1] While China is increasing its weight in the global economy, its economic achievements are mainly based on industrialisation. For example, more than half of the economic growth in 2011 came from "machinery, buildings and infrastructure".[2] Since the late 1980s, a

GDP (US$)

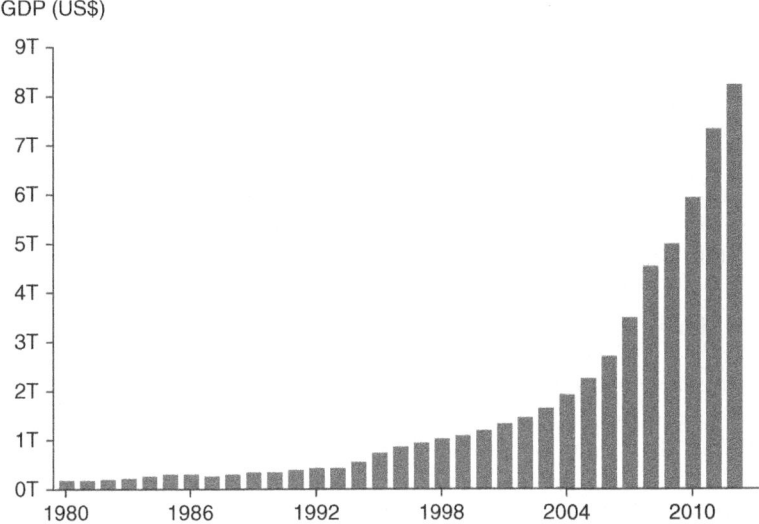

Figure 1.1 China's GDP from 1980–2012, according to the World Bank DataBank[3]

massive amount of foreign investment has poured into China. Foreign-owned factories have been launched, enjoying beneficial policies and cheap resources and labour. Giant transnational corporations, such as Sony and Unilever, have moved their production centres to China. As a result of this China has become one of the world's largest manufacturing centres. While the West has been transformed from an industrial to an information and knowledge economy in the post-Fordism era, China continues to produce hardware such as smartphones and computer accessories for the information and knowledge economy in the West.

After the 1980s, the Promethean belief and the Maoist attitude towards nature continued to function as the dominant principles that guided governments at all levels to develop local economies. The correlation between economic growth and political achievement further stimulated the enthusiasm of local governments in pursuing the growth of GDP. Local governments thus became obsessed with developing local econo-mies and attracting foreign investment with little concern for environ-mental protection.

Of course, against such a backdrop the lives of ordinary Chinese people have been highly improved. They are living a richer life nowa-days than in the 1970s, especially those who live in cities. People's life-styles have changed alongside an increase in their income and the fact that consumption is no longer suppressed as it was under Maoist rule (Ngai 2008; Tse, Belk et al. 1989). People are no longer proud of being proletarian and living a poverty-stricken life. Rather they are longing for a bourgeois lifestyle that encourages a desire for commodities and a pursuit of consumption. Consumption of middle-class products, such as cars and houses, and activities such as tourism, can now be afforded by ordinary people. That affordability symbolises a widespread consumerism in Chinese society (Ngai 2008) which, however, is not environmentally friendly.

1.2   The emergence of environmental problems and issues

All the developments discussed above have been accompanied by the deterioration of the environment (Zhang, Mol et al. 2007; Guo and Kassiola 2010; MacBean 2007) which puts "a better life" in question. While China becomes a factory for the world and more inclined to consumerism, its need for resources increases and the exploitation of nature is deepened. As an inevitable result, its environment is suffering from local resource scarcity and environmental pollution. The seven major environmental problems identified by Ma and Ortolano (Ma and Ortolano 2000: 2) in the early 21st century can be classified into these two types. While the scarcity of resource results from an unsustainable use of resources, intensive production without effective environmental protection leads to the pollution of the environment. Some environmental problems can be seen by the naked eye. Obvious river and air pollution and desertification, for example, can be sensed by humans. However, many other environmental problems, such as resource scarcity, underwater pollution and species extinction, are only brought to the attention of the public by being reported.

Global capital poured into China for a number of reasons. Cheap labour and extensive natural resources are attractive to global capital, eager as it is to accumulate profits. It was predicted that as a result of becoming a new resource provider for global capital, China would witness a new age of resource scarcity by the end of the 1990s (Walker 1996). In the mid-1990s China ran short of energy and grain, which had to be imported from abroad. Walker argued such a shortage resulted in an increase in food and energy prices as well as a shifting price mechanism, meanwhile posing a threat to China's dream of prosperity. Walker was far from being over-pessimistic. To take the energy shortage, for example, the situation in the new millennium is much worse than it was when his article was written. It has become widely accepted that the shortage of energy has become a crucial constraint on China's economic growth and even stability. The *Economist* revealed China was facing a severe problem in this respect in 2007 and it took over from the United States as the world's biggest energy consumer in 2011.[4] In the same year it was reported that China was suffering the worst energy shortage in years.[5]

In recent years it has been widely reported that cities that once enjoyed resource-based economic growth have now turned into "ghost cities" due to exhausting those resources. The state made a list of 44 such cities in 2008 and 2009. The economies of these cities were developed quickly in the 1990s, thanks to the exploitation of enormous natural resources such as coal, copper, oil and forests that were formed over millions of years as a gift of nature. Natural resources nevertheless declined quickly within a mere dozen years as a result of overuse (Wang and Guo 2011). Most of these cities are in the western regions, such as Datong in Shanxi, Baiying in Gansu, Panzhihua in Sichuan, E'er dou si in Inner Mongolia. They once had massive underground natural resources, such as coal and copper. Encouraged by the "Western Great Development" (*xibu dakaifa*) policies, these cities attracted a wide range of capital because of their massive local resources. This capital met with local officials eager to become rich. The economies in cities such as Datong and E'er dou si indeed grew very fast. However, after only a few years their natural resources were consumed and capital withdrew, leaving the devastated environment behind. Over-mining of underground natural resources led to geological disasters such as ground collapse, for example, in mining areas in Baiyin. In some areas, entire villages had to migrate to other places.[6] Identifiable social consequences of these environmental problems include the severe decline of local economies in general and particularly the collapse of local property and job markets.

A shortage of water is another major resource problem. While drought is continuously haunting Southern China (*huanan*), Northern China (*huabei*) is running out of underground water (Liu 1998). Many reasons account for China's water shortage, such as population-related and human-induced factors like overuse and the increasing demands caused by urbanisation and pollution and climate-related factors including droughts (Liu, Yu et al. 2001; Varis and Vakkilainen 2001). After assessing the risks associated with water shortages, scholars (e.g., Zhang and Chen 2009; Liu and Zheng 2002) even predict that Northern China will run short of water till 2014 or even 2050 despite all current governmental efforts. Scholars such as Dudgeon (1995) and McAllister, Craig et al. (2001) believe human interference, such as the South–North water transfer scheme (*nanshui beidiao*) and the construction of dams, will further disturb the ecological system and impact on biodiversity, and that this will lead to an even worse situation of water shortage. Conversely, other scholars (such as Liu and Zheng 2002) think that the potential damage caused by the South–North water transfer scheme can be considerably minimised by adopting new technologies.

Pollution is another major environmental problem. The over-reliance on coal as energy for industrial production is only one major cause for pollution in China (Smil 1996). Chemical, mining and manufacturing industries have posed enormous risks to the environment and there is a general lack of environmental protection measures and consciousness. In 2006, Pan Yue, the then deputy director of the National Environmental Protection Bureau, announced that some 45 percent of 7,555 chemical plants were sources of major environmental risks. Eighty-one percent of these factories were located in environmentally sensitive areas, such as near rivers and seas, and population-dense regions.[7] Industrial pollution accidents happened one after another across the country. In the single month of July 2010, for example, there were three major incidents in three regions: the Zijin mine pollution accident in Fujian Province, the raw petrol pollution caused by the explosion of a petrol transmission pipe in Dalian City, and the pollution of the Songhua River in Jilin Province by barrels of chemicals.

These environmental problems have posed great threats to animals as well as humans. Because of pollution and other human-induced environmental problems such as the construction of dams, the baiji, a freshwater dolphin, is presumed extinct. It was only found in the Yangtze River and no sighting has been documented since 2002. The porpoise is feared to be next. A national census published in 2013 reveals that there were more than 46,000 hydropower stations dividing China's rivers and lakes into fragmented pieces in 2011.[8] Hydropower stations have intensively impacted the environment in which a diversity of water animals live (Sun, Gong et al. 2013). For example, the Hekou hydropower station on the Yellow River has already led to the extinction of aquatic species in that river (Shen, Pan et al. 2009). Some species have become extinct or are facing extinction. Apart from aquatic species, the snub-nosed monkey is also nearly extinct due to the rapid and large-scale loss of forests. The Tibetan antelope faces a similar fate because of the rapid desertification of its habitat as well as humans' illegal animal skin and fur trade.

Although the environment is vulnerable to human intervention, nature takes revenge on humans through natural disasters and illnesses. Over the past decades, international and domestic news media (such as

Chu 1998) have frequently reported on natural disasters triggered by the construction of dams such as the Three Gorges Dam. The central government has even officially admitted for the first time that the Three Gorges Dam posed threats to the environment and ecological systems in the mid- and downstream areas of the Yangtze River in 2011.[9]

At stake in all this is whether modernisation can really bring a better quality of life for the Chinese people or if instead environmental problems are turning Chinese society into a risk society, in the term used by Beck (Beck 1992). This is because most of these environmental problems are human-induced and closely related to economic modernisation. However, they are problems that the current modern institutions are not able to handle effectively. Arguably it is far beyond the ability of industrial society to fix these problems. While some consequences of environmental risks have become visible already, many other consequences are dynamic and unpredictable and lie in the future. The health and safety of the Chinese people has already been threatened by environmental problems and disasters. Residents are shocked by the frequent appearance of abnormal, serious and even deadly environmental phenomena and climate changes, such as Cyanobacteria, earthquakes, ground collapses, smog, landslides, debris flows, sand storms, desertification and deforestation. These problems do not appear naturally, but result from human activities. Urban pollutants, for instance, were responsible for the environmental issues in the Taihu Lake (Qin, Xu et al. 2007).

Local residents in China have started to worry about the gravity of public health problems such as cancer, stone lung (silicosis) and lead poisoning, and the extinction of wild animals such as finless porpoises (Sanders 1999; Yang and Calhoun 2007; Liu and Diamond 2008; Fu, Zhuang et al. 2007; He 2009). Indeed, health problems and food security issues inter alia have become acute alongside the increasing severity of pollution. In February 2013 the existence of "cancer villages" caused by toxic chemical pollutants was for the first time admitted by the Environmental Protection Ministry.[10] People are worried about food being polluted or even poisoned by dangerous chemicals added to milk to make profits, as exemplified in the Baby Milk Powder Poison Scandal in 2008. Rice, vegetables and meat may also be contaminated by polluted water, air or soil. Experts and governments seem unable to offer effective solutions to the pressing situation regarding health and food security. Of course the awareness of these hazards comes both from what people witness with their own eyes and the media-constructed virtual reality about what others are experiencing. The warning bell was rung when these environmental problems became pervasive and made residents feel under environmental threat. This situation began in the 1990s and has become worse in the new century.

Meanwhile, environmental problems are economically expensive. Huge financial outlays are needed to fix environmental damage (Smil 1996). In 2006 the "China Green National Accounting Study Report 2004 (Public Version)"[11] was released by the National Environmental Protection Bureau and the National Statistics Bureau. According to this report, the financial loss (including damage caused by pollution, such as to fisheries, agriculture and health, and funds that are needed to clean the environment) caused by nationwide environmental pollution was 511.8 billion Chinese yuan, which was 3.05 percent of GDP in 2004. An estimated 287.9 billion Chinese yuan, around 1.80 percent 2004 GDP, was requested to fix environmental pollution in 2004.[12] The financial cost generated by environmental pollution has climbed to nearly 1,500 billion yuan in 2010, while the estimated financial investment for cleaning up the environmental pollution increased by 95.4% to 558.93 billion Yuan in 2010.[13]

In human history, as reported in Foster's seminal book *The Vulnerable Planet*, environmental changes and problems are often associated with, or lead to, changes in culture, the economy and even the whole civilisation (Foster 1999b). No political authority can afford to turn a blind eye to it. Aside from the accumulated doubts and complaints about environmental problems and food security felt in everyday life by the public, the severity of environmental problems in China has already led to a series of protests occurring frequently in recent years across the country. Environmental protests burst out as a physical expression of the public's discontent over environmental issues and government policies in this respect, as they may find it impossible to voice their interests in other ways. In doing so they push the authorities to tackle environmental issues and to make an appropriate and quick response. However, the situation is not that simple, as the authorities, who have to react to people's demands, are also under pressure to look after the interests of businesses for the sake of economic growth. This ambiguous attitude by the authorities throws up barriers for environmental protestors who look for solutions and environmental victims who seek justice. Many negotiations and compromises among governments, businesses and environmental victims are identifiable in this process.

Take 2012, for instance, which was a year of continuous environmental protests that received extensive attention in the public arena. Most of them involved large-scale mass protests in the streets opposing the establishment and operation of industrial factories and power plants.

Among others, in January there were local residents' street protests against the construction of a power plant in Haimen, Guangdong; in July, protests took place against the dumping of polluting toxic waste by a paper factory and its construction of pipes that will discharge sewage into the sea in Qidong, Jiangsu; and, separately, against the construction of a copper alloy factory in Shifang, Sichuan; and in July and August, against water pollution caused by the illegal mining, allegedly supported by local officials, of rare earth minerals.

A careful examination of all these events reveals the respective roles of various actors, with their different interests. Adapting the typology of Palmlund (Palmlund 1992), we can classify the actors into "environmental problems bearers", "environmental problems bearers' advocates", "environmental problems generators", "environmental problems solvers" and "environmental problems informers". Having different evaluations of risks, they act in particular ways that enable them to pursue their respective interests.

First of all, local residents played the roles of both "environmental problems bearers" and "environmental problems bearers' advocates". In these events, environmental issues touched a raw nerve among local residents expecting to live a happy life in a clean and safe environment. The protests originated from their fear of having their living environment polluted by these industrial projects. Second, economic actors – including both domestic and global actors such as transnational businesses and those at home – were "environmental problems generators". They often argued that the risks caused by industrial activities such as manufacturing and mining could be solved by advanced technologies. Third, local governments were also identified as "environmental problems generators", as they granted permission for the construction and operation of industrial projects or recognised and accredited the effectiveness of the environmental protection measures taken by them. In doing so, they tried to convince local residents that no harm would be caused by these projects. But after the outbreaks of these protests, local governments chose to give up global investors in the face of strong public disapproval of these factories. At that point they appeared as "environmental problems solvers" who try to demonstrate the ability and willingness to reduce potential risk. Finally, the news media played an ambiguous role in these events. While most of the local media stayed silent, national media and those in other regions heavily criticised the decisions by the authorities to build up these projects that had already caused damage to local environments or threatened to do so in the future. The media which gave voice to the protests successfully acted as "risk informers", raising these issues on the public agenda. These protests resulted in good outcomes, in which local governments promised to suspend the offending projects. However, it is impossible to predict if and when the local governments will restart these projects or move them to other geographical locations, as happened with the Xiamen PX protest in 2007, in which the PX project's location was merely shifted, finally landing in Zhangzhou City, another city in Fujiang Province. Therefore the success of "environmental problems bearers" and "environmental problems bearers' advocates" may only be limited to specific locales where these protests occur. Just like the anti-nuclear campaigns in the 1970s in the United States that were criticised for only caring about their backyard issues but lacking the intention to "democratise society" (Douglas and Wildavsky 1982: 141), protests of this kind are limited in effect and cannot fundamentally solve the environmental problems in China. Arguably, completely resolving these problems requires structural changes or large-scale shifts in central planning and modernisation strategies.

But these struggles among the varying parties in these protests vividly epitomise the social dynamics behind the rise of environmental problems and issues in China. Three key actors – the state, enterprises, and the public (local residents in the cases discussed above) – are playing their respective, but sometimes contrasting, roles in solving environmental problems. The state, encountering a dilemma, shoulders double responsibilities: achieving economic growth and dealing with the aftermath, in the form of environmental consequences. Global and domestic economic actors actively establish their businesses in China with the aim of making profits. The public – especially residents in localities where large industrial projects are based – perhaps welcomes these in the first instance, but later become more hesitant about them and even resist the operation of industrial plants in their "backyards". They want a better life. The news media somewhat mediate these struggles which lead to compromises among the actors.

1.3   The state: walking down a road of ecological modernisation?

It would be unfair to accuse the Chinese state of completely ignoring environmental problems. The modern state in China has responded to increasingly severe environmental problems by establishing a regulatory system with the goal of relieving them. The current environmental regulatory system is organised in a four-tier management system (corresponding to its administrative system) that has been established gradually over recent decades. Its appearance is at the centre of China's "reflexive modernisation". This term as used here differs from that defined by Beck, who refers to "reflexive modernisation" as a type of modernisation that is characteristic of post-industrial and risk societies. In a post-industrial and risk society, a reorganisation of industrial institutions is required to handle the risks that are generated and cannot be solved. This term, used to describe China's ecological modernisation, however, often conveys the meaning of reflecting on environmental governance and making changes accordingly in order to alleviate environmental problems, but meanwhile ensuring continuing economic growth. Therefore the purpose of China's "reflexive modernisation", which is analogous to "ecological modernisation", is still to boost GDP rather than to make fundamental changes.

According to Mol (2006), the Chinese state's true involvement in environmental protection dates from the 1970s. China attended the 1972 United Nations Conference on the Human Environment. The attendance facilitated the launch of the first National Environmental Protection Conference in 1973. At this conference a regulation outlining China's environmental protection strategy was issued for the first time.[14] The strategy was further articulated in the second conference in 1983, under the rubric of realising coherence among economic, social and environmental benefits[15] China's environmental governance began then, though taking an authoritarian approach. In 1974 China established a National Environmental Protection Agency and offices of this agency at lower administrative levels. The National Environmental Protection Law was enacted in 1979 and revised in 1989. Environmental protection was defined as a policy at the national level in 1984. A wide range of regulations and policies were passed in the 1980s which offered a sound foundation for China's environmental protection (Qu 2010).

The recent changes in environmental policies reflect the state's aim of sustainable development as a response to the long-lasting challenges the ecological system poses for its economic development (Zhang, Mol et al. 2007). In the 1990s the state sought solutions for environmental problems at the enterprise level, promoting clean energy and relevant policies to relieve the pressures on the environment (Qu 2010). A series of regulations and orders in the 21st century further signalled a shift towards sustainable development. Three of the most important regulations are the publication of the Ecological Environment Construction Plan in 1998, the Guideline for Ecological Environmental Protection, and China's National Guideline for Sustainable Development in 2000 (Zhang, Mol et al. 2007).

The state also hopes to get the public involved in environmental protection. In 2003 the Environmental Impacts Assessment Regulation was enacted to enforce sustainable development which encouraged the participation of the public. Later public hearings were introduced as part of the environmental management system which was a symbol of participatory environmental governance (Zhong and Mol 2008). Shi and Zhang (Shi and Zhang 2006: 271) summarise three features of the changes: "modernising the existing environmental regulatory networks, decentralising environmental policy and capacity, and including market and civil society actors and institutions in environmental governance".

The State Environmental Protection Administration (SEPA) launched a blitz on environmental protection from 2005–2007. The "China Modernization Report 2007: Study on Ecological Modernization", officially released in 2007, was seen as a sign of China's ecological modernisation. This report, which promoted the concept of a Second Modernisation, was seen as forming a good starting point for debate (Zhang, Mol et al. 2007).

Environmental or ecological modernisation has now become a useful term for policy-makers and social scientists in handling environmental conflicts, occupying a middle ground between the discourses of modern-isation (or industrialisation) and of de-modernisation (or de-industri-alisation) (Mol 2006). In the terms of environmental modernisation, modern institutions are to be restructured to meet environmental inter-ests (Mol 2006). This idea originates from modernisation in Western Europe. It sees environmental protection as having been increasingly compatible with economic growth (Hajer 1995) and views technology as key to achieving environmentally friendly modernisation (Christoff 1996). In this sense the term ecological modernisation is appropriate for describing China's attempt to deal with its environmental problems by restructuring its institutions (Mol 2006).

The term ecological modernisation nevertheless implies that, for the state, the fundamental solution to environmental degradation lies in an attempt to use new technologies and regulation to reduce damage to the environment, which suggests the supremacy of technology. This idea means overcoming environmental constraints for better economic development rather than slowing down economic development for the sake of protecting the environment. The underlying principle is to adopt advanced technologies or to explore new geographical areas in order to cope with the situation. This is still the Promethean and Maoist discourse that has prevailed in Chinese society for a long time. Therefore, as long as the national priority is placed on economic growth, it is not certain to what extent the environmental situation can be mitigated or whether it will become even worse.

In addition, three challenges to ecological modernisation are identifiable in China at present. The first is an opposition between those who prioritise environmental protection and those who prioritise economic growth. Based on the experience in Shanghai, Lo and others (Lo, Yip et al. 2000) examine the ways in which China's environmental regulation has been influenced by institutionally contextual factors and argue that its agency-dominated regulation is formal in requirements but informal in enforcement. Cooperation is lacking between environmental agencies and other government departments that are responsible for economic development (Lo and Leung 2000). A series of central government orders forced local governments to enact local environmental regulations in accordance with Beijing's desire to reduce pollution in the 1990s (Lo and Leung 2000). However, in reality local governments are among the major counter-forces opposing Beijing's interest in environmental protection (Wang 2008).

The second challenge is that when more and more global actors are involved as environmental problem generators, environmental protection becomes an issue that goes beyond national boundaries. The entry of China into the WTO has worsened the environmental challenges facing the country (Jahiel 2006). Along with the deepening of globalisation, global corporations and institutions have become part of China's ecological modernisation, partly accounting for the deterioration of the environment (Sonnenfeld and Mol 2002). Many of these multinational corporations have moved their industrial factories into China, which poses great threats to China's environment, especially when there is weak enforcement of environmental regulations (Economy 2006). Environmental problems that appear in the territory of China may therefore involve actors outside its borders. The deep involvement of multinational corporations in environmental destruction requires China to have an environmental governance that is in accordance with international environmental governance.

And finally, the solution to environmental problems often lies outside the purely environmental field. Instead, it often requires a profound institutional change in many other areas to solve such problems. The shortage of water in urban China, for example, requires a genuine institutional change across industries and cities to cope with it, which is certainly difficult to achieve (Nickum and Lee 2006). So are other environmental problems.

In 2013, after Xi Jinping came to power, new premier Li Keqiang made a speech in his first press conference which seemingly conveyed a new and exciting message on environmental governance to China and the wider world. According to a report from the Xinhua News Agency, Li said that "China cannot pursue economic growth at the expense of the environment, as the people would not be happy with such growth…. And governments should use an "iron fist" against environmental offences." He also pledged to raise the threshold of environmental protection.[16] Li's apparent determination has raised hopes for the resolution of environmental problems, especially pollution. Although it is a good sign, in view of the decentralised relationship between the central and local governments and the connection between local governments' governance assessments and local GDP growth, it is still too early to say how effective the new premier's environmental policies will be.

1.4   Enterprises: between maximising profits and being green

International and domestic enterprises are the main pillar of China's economy and industrialisation. According to the state Bureau for Industrial and Commercial Administration, by March 2013 there were approximately 13.75 million physical enterprises in China. Some 80% of these enterprises are private enterprises[17] and 4.4 million are enterprises with foreign investors.[18] Around 499 transnational enterprises (out of the world's top 500 enterprises) have made investments in China. Giant transnational firms such as Apple, Dell, Nike and Samsung have located their manufacturing operations in China, especially in the Zhujiang Delta, the Yangtze Delta and the Bo Sea Bay, which has turned China into the world's workshop. In 2012, manufacturing, mining, utility and building enterprises contributed over 45% of GDP.[19] These enterprises, however, are most likely to pose threats to the environment. Over more recent years there has been a tendency for these TNCs to leave China for other countries in Southeast Asia or elsewhere, where they can enjoy cheaper labour and more advantageous policies; for example, Adidas closed its operations in China in 2012.[20] Notwithstanding this tendency, there has been another wave of foreign investment which has reached a peak recently. This wave has been triggered by transnational petrochemical companies. A prominent example is Akzo Nobel, a Dutch company. Akzo Nobel's official website shows a long list of 59 firms and offices across China, including those based in Suzhou (Jiangsu Province), Ningbo (Zhejiang Province), Guangzhou, Shenzhen, Dongguan (Guangdong

Province), Langfang (Tianjin), Jiashan, Chengdu (Sichuan Province), Wuhan (Hubei Province), Shenyang Province, Songjiang (Shanghai), Kunming, Haerbing, Chongqing, Xi'an, Urumuqi and other cities. In 2013, Akzo Nobel announced its plan to invest €65 million more in its surface chemistry business in China and said that it will build a new alkoxylated equipment unit in Ningbo.[21] Therefore the potential for these enterprises to further damage the environment remains.

It is questionable to what extent the state can make enterprises effectively follow environmental regulations. For capitalist enterprises, their priority is, without doubt, making profits. Often environmental regulations and environmental assessment processes create obstacles for them in operating their businesses smoothly. For transnational corporations, for instance, the high environmental standards set by some countries require them to incur considerable costs before starting their operations. Apart from cheap raw materials and labour, the beneficial policies transnational corporation receive from local governments are one of the main reasons for them to move their production centres to China. Such policies include offering them low-priced, extensive land, beneficial taxation, and even low environmental assessment standards. It is China's local governments that have lowered the environmental entry barriers for these transnational corporations. Since the traffic light was set to green by local governments, a large number of industrial projects and plants have poured into China.

This phenomenon is extremely prominent in the petrochemical industry. Under the encouragement of the Petrochemical Industry Adjustment and Boost Plan[22] issued in 2009, local governments rushed to launch large-scale petrochemical projects in order to attract foreign investment and to achieve fast GDP growth. For example, while a campaign in Taiwan led to the cancellation of the construction of a petrochemical project in 2011,[23] by contrast seven Chinese provinces invited this project to come to their territories instead.[24] Lying behind the rush to attract petrochemical projects is local governments' ignorance of the needs of the environment and their lowering of environmental assessment standards. In 2011, the Ministry of Environmental Protection criticised 25 environmental assessment organisations from across the country for not providing honest information about the environmental impacts of petrochemical projects which they had assessed. These unqualified environmental assessment reports cleared obstacles and paved the way for local governments to invite large-scale petrochemical projects into their territories. Under such circumstances the petrochemical industry that has been encountering resistance from the public in other countries or regions, such as Taiwan, moved to mainland China in recent years.[25]

On the other hand, however, enterprises need to establish and maintain a green image in order to maximise profits. Enterprises need an environmentally friendly reputation and to meet both local and global standards for the sake of doing business well (Marqu, Zhang et al. 2011). This is partly because enormous economic costs will be generated if enterprises cause environmental disasters, such as pollution from an explosion, or if their business has triggered civil resistance. For those enterprises that are involved in or responsible for any environmental disasters, they face both public and economic pressures. Zijin Mining, for example, was fined 30 million Chinese yuan for its pollution incident in 2010. This company also suffered from a fall in its stock price after the national stock supervising committee started to investigate its operation after the incident.

Enterprises also need to show an environmentally friendly image for the sake of attracting customers. In the 21st century, more and more enterprises announced that they support environmental protection. Of course, it is arguable whether or not they really mean it and if their actions will match their statements. One never knows if they are really running a green business or merely engaged in an effort to greenwash their practices. Being green is also crucial for increasing the competitiveness of Chinese local enterprises in international markets. This maybe a precondition for them to offer outsourcing services or to become suppliers for transnational corporations that require them to be environmentally friendly. One effective thing for them to do is to achieve the green labelling certificates granted by the Chinese government. The China Green Labelling programmes started in 1993 when the Environmental Protection Ministry[26] issued the first green labelling regulation for products. It is not easy to assess quantitatively how effective these programmes are. A paper published in 1999 concluded that their effectiveness was "very limited" due to the inclusion of a small number of product categories (Zhao and Xia 1999). But certainly launching programmes like these can be seen as a determined attempt by the Chinese central government to regulate the practice of enterprises. Conversely, this can also be seen as an effective means to reduce green barriers for Chinese enterprises to enter international markets and to influence customers' purchasing behaviours if their products can be legitimately "green labelled".

Apart from looking for political protection from local governments, enterprises seek to maintain their green image through the adoption of advanced technologies. It is common to come across attempts by enterprises to establish a logic that advanced technologies can help remove environmental risks in order to show their green credentials. Take Shijiazhuang Medicine Manufacturing Group, for instance. In 2012 it was proud of its 790 million yuan advanced clean technologies, guaranteeing that no pollution would be caused by its production process. In the same year, it even won an 8.5 million yuan award from the Industrial and Information Ministry and a 6.75 million yuan award from Hebei Province.[27] A single click on the websites of any chemical factories or other manufacturing companies will reveal similar boasts of this kind. However, the claims of the effectiveness of advanced technologies to protect the environment really need hard evidence and time to assess whether they are simply empty boasts.

1.5 Social class, environmental movements and Chinese society

To be rich or to have a clean environment, that is the question. In general, Chinese people face a dilemma of this kind. At the beginning of the opening-up phase of the economic reforms in the 1980s, the discourse of becoming rich was encouraged by Deng Xiaoping's Cat theory[28] and became pervasive in the whole society. At first people welcomed the industrialisation and marketisation of the economy and the urbanisation of remote areas because they brought jobs and opportunities to be richer. However, after two decades, when the deterioration in the environment became apparent and serious health problems emerged, people started to realise the importance of protecting the environment, though for the most part it was too late to stop these environmental problems from occurring. Moreover, such an awareness of environmental issues never means people actually prioritise environmental protection over the need to become richer and to have a luxury consumer lifestyle.

At first glance, environmental problems are related to social class as usually it is people from the lower class that are victimised by environmental problems. Environmental deterioration has resulted in a large number of environmental victims and environmental injustices. A question that arises is who bears the burden of environmental problems and suffers the consequences of environmental deterioration in this country. Liu (Liu 2010) has found a geographical pattern of environmental injustice in terms of the appearance of cancer villages, that is, cancer villages are more likely to appear in less developed and poorer areas than in more developed and richer areas. The poor become poorer due to illnesses caused by environmental problems. In addition, class and economic status do matter in this respect. A study by Ma (Ma 2010) reveals environmental inequality in China: rural residents and those who have migrated from rural to urban areas suffer most from environmental problems. This means a relationship has developed between environmental problems and social hierarchy. Lower-class people are more likely to become environmental victims than those of higher classes, while the latter tend to be environmental risk generators. Thus, emerging concurrently with environmental problems is the restructuring of social classes during the modernisation process in China. Different relationships to environmental problems can even differentiate one social class from another. The amount of social resources that people possess influences to a great extent the relationship they have to environmental problems. Natural resources, such as land, are the first sacrifice peasants have made in China's urbanisation and West Development processes. Peasants have to work in factories after they have lost what they depended on for their living. Environmental problems thus moved from urban cities to villages and from the east to the west. Low-income workers and landless peasants are turned into environmental victims. Both in the pollution disaster in Zijin event or the case in Daye Hubei where resources were quickly exhausted, for instance, it was local fishermen, peasants, mine workers and local residents who became environmental victims. This reality may harm the interest of the majority of China's population, who are peasants and workers, and thus shatter the foundation for the legitimacy of the CCP.

Chinese society has witnessed the rapid rise of the "middle class", comprising civil officials, managers, intellectuals and professionals (Lu 2013). This group of people enjoys decent standards of material life and can afford expensive goods, such as housing and cars. For example, the rise of the middle class has boosted car sales in China, which is not environmentally friendly at all.[29] Economic and political elites, such as capitalists and politicians who possess power and large social capital, benefit most from economic modernisation. Some groups of the elite issue and implement both economic policies and environmental regulations. Other groups of them directly exploit resources to make profits for their enterprises. These groups even enjoy "special supplies" (*tegong*) of resources and products, such as food, wine, medicines, water and clothes, which are especially made, grown and supplied to them so that they do not need to worry about contamination. The existence of such "special supplies" gives them the false sense that they can be exempt from environmental problems.

However, environmental problems may, in fact, affect all classes, although this takes time for people to realise. All social classes are ultimately equal in terms of environmental problems. Foster has examined global capital's exploitation of nature and argued that those in the Third World, such as "the low-lying delta of the Pearl River and the Guangdong industrial region from Shenzhen to Guangzhou" in China, are the most oppressed by the capitalist mode of production and have nothing to lose. Therefore they are "environmental proletariats" (Foster 2010; Foster, Clark et al. 2010). This argument is very interesting as it treats people living in China as equally suffering environmental problems while acknowledging the existence of environmental inequality on the global level. It is true that no one can truly escape the influences of the environment. All of us are environmental proletariats. Both middle- and upper-class people can be environmental victims when their living space is intruded on by the construction of chemical factories or rubbish-burning projects or when they suffer pollution. These elite social groups cannot entirely escape the fate of being victimised by the deterioration of the environment unless they wear oxygen masks to prevent themselves from breathing polluted air or even leaving for a cleaner place to live. The differences between lower- and higher-class people thus lie in whether or not victims can leave the damaged environment for a better place and whether or not victims can seek environmental justice.

Without intending to treat them as a homogeneous group, it would be fair to say that generally ordinary people pay the price for the deterioration in the environment. The cost comes in multiple forms. It may take the form of sacrificing people's culture and communities to modernisation. For example, people who used to live in the three-strait area and were forced to leave and migrate to other areas were reportedly called "three-strait refugees" who lost their homes, culture and communities to the construction of the giant dam. The cost may also be in the health, safety or financial losses of people who are victimised by a dangerous environment. Contaminated villages where many residents have contracted cancer have been labelled "cancer villages" by the mainstream media (Liu 2010). People living near large petrochemical projects, such as residents of Dalian, are continuously threatened by the possibility of an explosion. Even a single industrial incident causes huge financial losses for fishery or agrarian owners.

All this suffering and all these losses cannot be fully relieved and reimbursed for four reasons. The first is the difficulty in identifying individuals or institutions that should bear the responsibility for the problems which caused the loss and suffering. It is not only that environmental victims are denied justice, but also that they do not always know from whom they should seek justice. The basic idea underlying China's environmental reimbursement system is a principle of "who causes should reimburse (*shei shunhai shei buchang*)". This principle was set by the 1996 regulation "the state Council's decision about several problems of environmental protection". The responsibilities for restoring the environment and reimbursing environmental victims lie with the individuals and organisations that have caused the environmental problems, the same as the principle of "the polluter pays", widely recognised in OECD countries. However, it is often difficult to identify those who should be responsible for environmental problems, due to the unclear boundaries of responsibility. Much environmental damage is indirect (Goldman 2006/2007). China's administrative structure also increases the difficulties of this, as the structure divides administrative responsibilities according to regions and therefore separates regional responsibilities. On top of these obstacles, the lack of efficient regulation enforcement leads to difficulty in bringing environmental damage litigation (Wang 2007). A deficit in legitimate regulation enforcement makes the situation worse, on top of the question of national and local environmental regulations by the State Environmental Protection Agency (SEPA) and its local bureaus (Goldman 2006/2007).

An obvious example of this is "the flow of pollution", that is, downstream pollution caused by upstream activities. In 2010 fish farms downstream in Guangdong Province suffered enormous economic losses from pollution caused by the Zijing copper mining company in Fujian Province. But it was difficult to collect evidence to prove that the costs incurred by the victims were caused by the Zijing pollution incident[30] and therefore they were unable to receive compensation for their losses. It required negotiation and collaboration between the two provinces.

The second reason for the deficit in enforcement and reimbursement of losses is the intimate collaboration between local governments and environmental –problem generators. Local governments, who often favour local economic growth over environmental protection, tend to protect local enterprises from being punished for causing damage to the environment (Wang 2007). It is not rare at all to see industrial plants that have passed local governments' environmental impact assessment but are later revealed as polluters. For example, even after the Zijing mining incident, the water quality of the Ting River, that had been severely polluted, was still confirmed by the local government as having met environmental standards.[31] Due to corruption or the lack of financial resources, local governments have been found to be unwilling to enforce environmental regulations against local enterprises that contribute to the growth of the local economy (Goldman 2006/2007).

Thirdly, the responsibility placed on the victims in collecting and presenting evidence of the causes of their losses creates difficulties for achieving justice. Their capacity to fulfil this responsibility is constrained by the lack of resources and often by their background. This point has been mentioned repeatedly in the in-depth interviews carried out for this book with investigative journalists about their views of the barriers to achieving environmental justice. According to these journalists, these victims often are from lower social classes and have low-level literacy skills and financial resources which limit their ability to achieve environmental justice.

And, finally, the overall national priority for economic growth generates general difficulties in mitigating the situation. As long as GDP is still set as a key measure for governmental achievement, environmental protection will never become a real priority on a par with economic growth. The overall GDP guideline, for example, will only lead to moving heavy-polluting plants from developed to less developed regions, a phenomenon known as "pollution migration". With the priority on economic growth, such plants will never be completely closed, which is the only thing that can truly eliminate the cause of their pollution. As Wang argues, it is necessary to replace the traditional GDP measure with green GDP to reduce the ecological costs caused by modernisation (Wang 2007).

Proactive participation of citizens and consumers is needed in the current social and economic context (Martens 2006). The demands from people for a clean environment to live a good life constitute a bottom-up pressure on the leadership. Grass-roots resistance has been initiated by NGOs and citizen activism against top-down political and corporate power. A wide range of environmental movements have been identified during these years, signalling possible changes in the relationship between the state and society in China. The state–society relationship in China has been characterised by "strong state and weak society" since the establishment of the PRC. Authoritarianism is practiced in the day-to-day lives of Chinese people, ranging from the Confucian tradition to the more recent state-driven economic marketisation and from collectivism to proven loyalty to the Party. Despite this, society is believed to have obtained certain levels of autonomy because of China's opening up to the world, the speedy growth in individuals' wealth, improved living standards and the emergence of a considerable middle class.

The appearance of social movements that oppose the decisions of the state on certain issues and try to change the status quo is a sign of that autonomy. Although movements with the aim of advocating political reform have almost automatically disappeared from the public arena since 1989, environmental movements, *inter alia*, are most prominent, involving the participation of environmental NGOs, activists and even ordinary citizens. Environmental and non-governmental organisations are the first type of environmental protection force (Yang 2005). NGOs that are funded either by domestic sources or international investment have flourished in China since the 1980s. According to the China Development Brief website,[32] the number of domestic and international NGOs has increased to 268 in 2012 (with 225 NGOs on environmental and animal protection and 43 on climate change). The civil society role of environmental NGOs has been prominent since their emergence in China in the mid-1990s. Especially in the earlier years, before the 21st century, domestic NGOs played a leading role in launching environmental movements, as exemplified in the anti-deforestation (snubbed-nose monkey protection) campaign and the campaign to protect the Tibetan antelope. In the new century, international NGOs such as Greenpeace are more actively participating in environmental movements. The campaign against the Jinguang Group's deforestation has lasted for more than five years and has gradually achieved some progress recently.

The relationship between the media and NGOs has been an interesting and intricate one in China. NGOs, such as The Friend of Nature, and more recently China Dialogue, act as advocates for environmental reporting. Activities such as "Best Environmental Investigative Journalists" from 2011–2014, organised by China Dialogue, environmental and health reporting workshops held by Science Squirrel Association (*kexuesongshuhui*), and "Ten Best Environmental Journalists" organised by The Friend of Nature from 1996–1999 help in that role. Moreover, as early as the start of the 21st century, scholars identified a close link between media/journalists and NGOs (Mol 2009). It is not unusual to see professional journalists move to work for NGOs. The blurred boundaries between journalists and NGO activists can lead to a lack of clear differentiation between the two occupational groups. Journalist-motivated NGO organisations, such as Green Earth and Green Beagle, further connect the fields of journalism and NGOs and the public, extending the possibility of interacting among the three. Green Earth, to take an example, whose founder is veteran journalist Wang Yongchen, and *China Youth* regularly organise journalist salons. Environmental journalists such as Zhang Ke from *First Financial and Economic News* regularly attend the seminars, either as speakers and organisers or as members of the audience. The topics discussed in journalist salons often set the agendas for media. The public are welcome to attend the salons which often host experts' speeches on environmental issues.[33]

The case of Apple in 2011 is a prominent example of the collaboration between journalists and NGO activists. In August 2011, five environmental NGOs published a report accusing the company of severely polluting the environment in China. Some of these NGOs were founded by ex-journalists, for example, Green Beagle was founded by Yongfeng Feng from *Guangming Daily* (*guangming ribao*). The campaign received extensive media coverage and exposed the terrible environmental problems and working conditions that threaten workers' health in Apple suppliers' China-based factories, which might be more properly described as sweatshops. The issue received a great deal of attention and provoked considerable public anger. In November 2011, Apple finally admitted that its 15 suppliers had caused the pollution and agreed to raise the environmental standards for selecting its suppliers.

Citizen activists and ordinary citizens have both joined the camp of environmental movements. Protests or campaigns ranging from the anti-rubbish-burning project campaign in Guangdong to the anti-chemical (PX) project protests in Xiamen and Hunan represent local residents' resistance against these industrial projects. Middle-class social groups are often a particular pillar of campaigns of this kind. Recent social resistance movements against the launch or operation of industrial projects have prompted scholars such as Yang and Calhoun (2007) and Ho (2001) to believe that a green public sphere has appeared in China. Yang studied environmental NGOs and other communities and argued that Internet-based environmental groups represent public participation from the bottom up with the aim of dealing with environmental issues (Yang 2003c). More recently, especially facilitated by the wide application of social media tools such as Weibo and Webchat, the agitation from the bottom up calling for environmental problems, such as air and water pollution, to be resolved, has pressurised Beijing into making changes in the relevant policies.

The case of PM2.5 – a measure of tiny particles of pollution in the air which can be especially hazardous to health as they can be absorbed deep into the respiratory system – is a successful example of this. At the end of 2011, while Beijing was declaring that its air quality was fine, the American embassy in Beijing published its own PM2.5 data, based on hourly monitoring, on its Twitter feed (@beijingair), which was very different from that publicised by the Chinese government and suggested there was severe air pollution in Beijing.[34] The difference triggered heated online discussions that questioned the credibility of official pollution statistics. Some ordinary citizens, such as Pan Shiyi,[35] who could afford the relevant equipment, even started to monitor PM2.5 data by themselves and then published the data on their Weibo accounts. Overnight it appeared that everybody was talking about PM2.5. Despite protesting against the US embassy's monitoring of PM2.5,[36] in 2012 and 2013 the Chinese government responded by setting up PM2.5 monitoring stations in around 200 regions and cities and promising to publish PM2.5 statistics. This case suggests the important roles played by different actors – foreign interference and pressure, media technologies and people's environmental consciousness – in influencing Beijing's environmental policies, as well as in raising public awareness of environmental problems.

However, the public's efforts in agitating for a better environment are limited by two factors. To pay attention only to what happens in one's own backyard tends to prevent environmental movements from impacting on environmental policies in general. It would be difficult for citizens' protests to move from focusing on individual incidents to becoming large-scale social movements aiming at democratisation. On the other hand, Chinese people – especially upper and middle class Chinese – are most often consumers first and foremost rather than pure citizens. Consumer culture plays an important role in their lives. The accelerating pace of consumption is indeed the origin of many environmental problems, such as pollution and resource scarcity. It is far from being a civil society and environmental public sphere in a true sense.

1.6 Mediating the risks of environmental problems: how far can investigative journalism go?

Generally speaking, the news media are expected to play a crucial role in environmental governance and protection (Stuart, Barbara et al. 2000). The media can expose and disseminate information to the public, influencing the latter's perception of the state of the environment. Chinese traditional media, however, operate in an "information poor" social environment (Mol 2009) resulting from authoritarian control that treats the mass media as an ideological apparatus. Somewhat surprisingly, the news media have given intensive attention to environmental problems for several decades. This is thought to have originated from a top-down push from the state rather than merely being initiated by journalists (Wiest 2001). Environmental reporting was ignited along with the launch of the environmental monitoring and protection system. Environmental reporting thus functions as part of the environmental monitoring system and partly plays the role of environmental propaganda in China (Wiest 2001; Mol 2009).

As discussed above, in 1992 China agreed to abide by Agenda 21 of the United Nations with regard to sustainable development. A series of programs and actions have been put forward by the Environmental Protection Committee of the National People's Congress (NPC) in order to raise public awareness across the country. A massive reporting campaign was launched to organise media reports on environmental issues (Wiest 2001). China Environment Centennial Journey (*zhonghua huanbao shijixing*), starting in 1993, for instance, was a major environmental reporting campaign of this kind (Qie 2007; Yang and Calhoun 2007).[37]

More recently, mainstream media have been given formal entitlements to report on environmental information by relevant regulations and laws. For example, SEPA's State of the Environment Report issued in 2000 and the Greenwatch program later legitimised the publication of the pollution records of industrial plants by state-owned media (Mol 2009). This has proved effective at local levels, given the particular media control structure in which governments can only control the news media at the corresponding administrative level and operate within their administrative territories. Notwithstanding the freedom achieved for reporting on local environmental problems, more generally the transparency of environmental governance is limited by contextual factors such as the control of information by the government and the priority accorded to economic growth (Wang 2005) and the intimate relationship between local businesses and governments. Other regulations and laws, such as the 2002 Clean Production Promotion Act and Regulations on Open Government Information, have given the media the right to publicise and disclose governmental environment-related data (Guo 2005; Mol 2009).

The central government's approval and support have encouraged the media, especially the national media, to report on environmental issues, often those that are happening at the local level. There was a proliferation of environmental reports in the 1990s. From 1993 to 2008 more than 80,000 journalists attended China Environment Centennial Journey. They published or broadcast more than two million reports in traditional media.[38] As early as about ten years ago, journalists already remarked that they have more autonomy in reporting environmental topics than other socio-political issues (Wiest 2001). This view was confirmed and reiterated in my interviews with journalists from 2011 to 2013.

Apart from the impetus towards environmental reporting from above and the associated autonomy, expectations from the audience also pushed the media to increase coverage of environmental issues. There has been a dramatic increase in the public's concerns and even complaints about environmental problems. From 2001 to 2010 the number of petitioners (including those writing letters and making visits collectively or individually) to the Ministry of Environmental Protection (previously SEPA) increased over 545,000 to 802,000.[39] Despite the opinions voiced by experts that PX projects are environmentally friendly, there was one protest after another in different places against the launch of PX projects, and this demonstrates well the conflict between government's priority for economic development and ordinary people's concerns about environmental risks. This is exemplified in the very recent Guangdong Maoming Walking Protest against a PX project in 2014. Reporting on environmental issues shows the attention conventional media has paid to the needs of the people, as well as to the requirements of the market, as this reporting also helps the media to increase financial revenues.

Chinese media, like media elsewhere, are sites of power struggles. Various parties, including, among others, environmental victims, NGOs, governments and enterprises, have their own interests with regard to environmental issues. After several decades of environmental reporting, the topics of such reports have extended from the activities of relevant governments and officials to those concerned with ordinary people's attempts to assert their rights, such as the pursuit by environmental victims for justice and compensation. Given the traditional intellectual function of Chinese journalists, environmental victims naturally seek the help of journalists, expecting them to report their unhappy and even miserable experiences so that they can obtain justice. However, the media and journalists produce reports according to news value and professional criteria rather than sympathy for victims or the sense of mission they might have. It is true that under some circumstances media coverage does help solve problems that ordinary people are unable to solve otherwise, but it is unlikely to function as a medicine that can cure all diseases. Therefore there is a gap between the public's expectations and the function of environmental reporting.

In spite of all these positive examples of the media taking up environmental issues, their capacity to do so is limited by the unique relationship the Chinese media have to the authoritarian state and the ruling Party. Chinese media not only operate under the control of the party-state to serve the interests of the party-state, but also operate in the market to meet their commercial needs. Some scholars argue that the

Chinese media serves mainly to polish the image of the ruling party and legitimise the policies and rule of the Party (Lee, He et al. 2006). Despite being seen as less politically sensitive than many other social issues, environmental reporting is still constrained by the efforts of the state and other social actors to manage and manipulate media coverage. First of all, there are still forbidden topics in this area, such as problems with the Three Gorges or planned dams, and reporting bans still exist. Environmental audit reports, for instance, mostly do not get published in the mainstream media (Mol 2009). In addition, regardless of the supportive attitude of the central government, local governments disapprove of the disclosure of problems caused by industrial plants in media coverage which, they believe, may impact on local GDP growth (according to the interviews with journalists).

As discussed above, enterprises have now realised the importance of having an environmentally friendly public image for their business. Thus they try to establish and maintain a good relationship with media through offering advertising or other types of financial subsidies to media outlets (according to the interviews with journalists). This results in a contradictory relationship between the media and enterprises. On the one hand, enterprises are often the target of criticism by the mass media, especially for planning or operating plants that have generated or may generate damage to the environment. On the other, however, they are sponsors for the news media and even for environmental movement activities organised by the news media, such as *Southern Weekend*'s annual Green Media Development Program (*lvsechuanmeichujinjihua*).

Apart from all that has been outlined above, a proliferation of green websites in the last decade or so has brought opportunities to all parties. The decentralised technological features of the Internet make possible the circumventing of the dominance of the traditional news media in environmental reporting. On the other hand, user-generated content on the Internet provides news leads for journalists and amplifies the effects of environmental reports on many occasions.

It is questionable how far the Chinese media can go in reporting environmental problems under these circumstances, facing the social dynamics described. After all, how the news media represent environmental problems in their coverage is even more important than whether such representation exists. Interesting questions include (but are not limited to): What kind of discourse surrounding environmental problems has been constructed? How do environmental investigative journalists do their job? How do these discourses impact on China's modernisation? What sort of cultural politics have Chinese journalists adopted, as discussed in the work of Cottle (Cottle 2000)? Have the news media paid attention to a variety of claim-makers? Do they continue to give a voice to elites or experts as "primary definers" of environmental problems? Do they give a voice to grass roots or lay people? What types of environmentalist discourse prevail in Chinese society? Can discourses of this kind serve as a counter-hegemonic force and create a new hegemony that is opposite to the hegemony of economic modernisation? These are questions to be discussed later in the book.

1.7 Conclusion

The discussions about environmental problems, their relationship to modernisation and the responses of various social actors in this chapter provide a detailed base for the issues that will be explored in the rest of the book. To understand how the environmental situation has come about in China and the role the news media play in the process, it is necessary to consider the mutual influence between the environment and society. The case of China echoes the argument of environmental sociologists that environmental problems are about society rather than about nature (Hannigan 2006). What is evidenced in the physical environment actually results from an interaction of the environment with dynamics and power struggles in social and political environments. This point is very true in the case of China, as what accounts for China's environmental problems are politics and economics rather than nature itself. The integrated impact of political and economic dynamics on the environment has become prominent after more than half a century of modernisation.

The discussion in this chapter shows that the intertwined impacts of social dynamics on the environment are rooted in two contradictions in China's social logics in the relationship between nature and development. These two paradoxes underpin the pressing environmental situation in China. One is associated with an intrinsic paradox in Chinese society's longing for a better quality of life: a top-down pursuit of rapid economic development and a bottom-up urge for a clean and safe environment in which to live. The former is a form of Promethean discourse that believes the state and elites are able to fix the problems of resource scarcity, if there are any, and thus it justifies the capitalist mode of production and industrialisation (Dryzek 2005; Foster 1999b; Hannigan 2006). This discourse is promoted by the state and its associated economic and political blocs and is incorporated into the national priority for economic growth. The latter, which is a populist discourse of environmental risk, however, is initiated by ordinary people and grass-roots organisations as a response to the environmental crisis. Ordinary people, who often pay the price for environmental deterioration, strive for a clean and healthy environment. As their physical environment has fallen victim to economic modernisation, directly threatening their quality of life, they are eager to steer the attention of the top from economic growth to environmental problems. In doing so, they seek a way to fix environmental problems, through pressing the Chinese government to reflect on its economic modernisation plans and issue environmentally friendly public policies on modernisation.

The other contradiction originates from what Schnaiberg calls an innately conflicting relationship between "the treadmill of production" and the need for environmental protection (Schnaiberg 1980). The capitalist drive for economic growth demands no abandonment of production and exploitation of the environment at all. Instead, they place reliance on the opening up of new areas, the changing of geographical locations, the issuing of new policies and regulations, or the employment of new technologies to reduce the damage economic growth may inflict on the environment and to relieve environmental deterioration. That is why ecological modernisation is widely promoted.

These two pairs of social logics underpin China's modernisation process as well as driving power struggles among a diversity of interest groups when dealing with environmental problems. The media play a vital role in mediating among these different parties, channelling their respective demands, on the one hand, and becoming the site over which these interest groups struggle for control on the other. Given its state ownership and political control, it is uncertain how far the Chinese media can go in this respect. The book aims to offer some answers to this question by examining to what extent investigative journalism can facilitate the formation of discourses that embody environmentalism and can therefore act as a counter-hegemonic force. The next chapter will examine the development of environmental problems and the reporting of them over the past 20 years.

2
Twenty Years of Environmental Investigative Reporting: Agendas, Social Interests and Voices

Over the past 20 years, environmental problems have developed into a major topic for investigative journalism. This development was against the backdrop of continuing economic modernisation and generally emerging environmental problems and movements. During this period, China has experienced four "five-year plans" (*wunianjihua*)[1] and three generations of leadership. Meanwhile, state capitalism has been systematically established, expanded rapidly and spread from the centre (the eastern and southern coastal cities) to the peripheries (the western, northern and middle inland areas). An inflow of global capital into China has been observed in all sorts of areas, ranging from manufacturing and agriculture to mining. As a result, Chinese society has undergone a dramatic transformation, one aspect of which is the forming of social groups that have their respective interests and distinct value systems. These value systems, diverging from the main values promoted by the Party-state, underlie increasingly active citizen participation and lobbying to change government policies and decision-making on environmental issues.

Environmental investigative reporting in the 20 years from the late 1990s to the present has constructed nine agendas regarding environmental problems, depicting an appalling picture of environmental destruction that suggests the environment is at risk from the practice of modernisation. A close examination of the development of these agendas brings to light three interesting findings concerning the relationship of investigative journalism to social institutions and values. First, the relationship between investigative journalism and environmental non-government organisations (ENGOs) has changed over time. Investigative journalism has taken over from the ENGOs in setting and defining agendas, whereas the ENGOs' leading role in sparking environmental movements in the 1990s has declined and has now been more or less reduced to acting as news sources for investigative reports.

Second, overall the nine agendas reflect different social interests and voices and expose environmental hazards in diverse ways. Some agendas, such as pollution and resource exhaustion, cater to the interests of the state and the market more than others, while agendas like anti-dam construction and anti-artificial island/land construction strongly reflect critical voices which are opposing the state's policy, instead of being inimical to political and commercial interests. Yet other agendas, such as cancer villages, voice the interests of the underprivileged and enable them to draw the attention of society to their misery, caused by the damaged environment.

Third, and largely as a result of the second point, the nine agendas indicate the diverse effects of environmental investigative reports and an ostensibly self-contradictory function of investigative journalism. On the one hand, to our surprise, investigative journalism functions as an "ideological state apparatus" as theorised by Althusser (Althusser 1971) that helps to consolidate the control and rule of the state, reproduce existing capitalist relations and create consent in society. There is coherence between environmental investigative reports and the interests of the state and the market. This coherence implies the ideological function of investigative journalism as it complies with dominant political and commercial interests. On the other hand, investigative journalism has manufactured – or mediated – dissent about modernisation through its construction of a discourse about environmental problems produced by modernisation. Dissent reflects and facilitates contestation among different social values prevailing in Chinese society. This seeming self-contradiction in the role of investigative journalism reveals the function of journalism to serve multiple and contesting social values, among which the interests of the state and the market are the major, but not the only, examples. This can also be seen as a case in which journalism benefits from the support of the state, but meanwhile grasps the chance to struggle free from the control of the state and satisfy multiple interests in society.

2.1 Environmental investigative journalism: its development and relationship to ENGOs

Although environmental reporting had already appeared in China in the 1980s (Zhang 2007), environmental problems did not start being a topic for investigative journalism until the late 1990s. Environmental investigative reporting benefitted from the introduction of top-down administrative requirements to report on environmental problems for the purpose of educating the public and correcting the wrongdoing of individuals who might damage the environment. By the 1990s, China had already walked down the road of modernisation for nearly half a century. The market reforms had been imposed for more than 20 years. The long-established philosophy of "men conquer nature", and the priority accorded to economic development, have driven governments – at all administrative levels – and individuals to pursue economic growth without paying much essential attention to environmental protection. By the late 1990s, people began to be aware of the damage that had been done to the environment, impairing the legitimacy of the ruling CCP (Holbig and Gilley 2010). The central government began to realise the importance of environmental protection and thus launched a top-down environmental propaganda campaign. As a result, investigative journalists started reporting on environmental problems at a time when concern for the environment began to emerge in society as an outcome of the top leadership's awareness of the conflict between development and nature.

As introduced in the previous chapter, in 1993 the propaganda programme called "China Environment Centennial Journey" officially began, organised by the National People's Congress and 14 other central departments including the Central Propaganda Department (*zhongxuanbu*), the Agricultural Ministry and the Environmental Protection Ministry. With the aim of increasing public awareness of environmental protection, this programme was state-initiated and involved news media across China in reporting on local activities that were either good or bad for the environment. News media at the central level, such as the *People's Daily*, CCTV and the Xinhua News Agency, as well as those at the local level assigned journalists to participate in the reporting programme, travelling around the country to investigate and report on both the achievements of environmental protection and environmental problems.

This campaign/programme continued for the next 20 years. Every year the programme officially sets up a reporting theme for journalists. From 2000 to 2013 there are some 2,344 reports contained in Chinese newspapers that are related to this tour.[2] Most of the reports are about pollution – either pollution problems or effective solutions to them. Occasionally, some reports are critical of particular cities or enterprises for overlooking environmental protection and other reports directly praise the achievements of these cities and enterprises after having been criticised.

The start of this propaganda campaign passes a strong message to both the news media and society that environmental protection is at the forefront of the central government's attention. The central government expects news media to take on this public opinion supervision role (*yulunjiandu*) on the premise of adhering to the main theme (*zhuxuanlv*) it has set. This programme reminds local governments of their duty to look after the environment while developing economies and also justifies reporting on environmental problems and issues. This is indeed a symbol of the launch of an internal surveillance system for monitoring environmental problems.

The initiative of the state matches and encourages the aspiration of environmental activists to promote environmental protection from the bottom-up. Environmental NGOs started emerging in the mid-1990s (Wu 2003). Some members of the elite, such as journalists, who possess media and political resources, led these ENGOs. They launched a series of campaigns in which a range of social actors, ranging from ordinary citizens, news media and government officials, played their respective parts. The establishing of environmental NGOs such as Friend of Nature (*ziranzhi you*) in 1994, Green Home (*lv jiayuan*) and Beijing Earth Village (*beijingdiqiucun*) in 1996 marks the start of the bottom-up initiative that has driven environmental movements since then. These civil organisations endeavour to protect the environment and to advance environmental awareness in Chinese society.

Thanks to the activity of environmental NGOs, the environmental movement in China reached a peak by the end of the 1990s and the early 21st century. Important environmental campaigns in the 1990s included anti-deforestation campaigns to protect snub-nosed monkeys in Yunnan as well as campaigns against the construction of hydroelectric projects and dams. Environmental movements in the 1990s were elite-oriented, drawing on the alliance between news media and NGOs. Environmental campaigns were largely limited to sending written appeals to political authorities or officials at higher administrative levels, publishing critical reports in the media, launching exhibitions and so forth. The extensive participation of ordinary people was rare. Therefore, the 1990s' environmental movements are quite distinct from those in the 21st century, in which environmental protests feature prominently.

Investigative journalism emerged in China around the mid-1990s, mainly focusing on social issues. The attention of investigative journalism to environmental issues in the late 1990s was closely linked to the active role of ENGOs. The prominent collaboration between investigative journalism and ENGOs in the 1990s is best exemplified in two important and influential environmental movements: anti-deforestation (involving threats to the species *Rhinopithecus bieti*, also known as the snub-nosed monkey) and Tibetan antelope protection (anti-illegal hunting and trading in Tibetan antelopes). In both cases we can see ENGOs taking the leading role in agenda-setting and getting the message across to high-ranking officials. Investigative reports in the news media only played a complementary role, helping ENGOs to get the messages out. But investigative reports enlarged the influence of ENGOs' campaigns in the late 1990s. These campaigns would not have been so influential without investigative reporting. The effectiveness of ENGOs and investigative reporting is somewhat related to social dynamics, such as the involvement of the central government in snub-nosed monkey protection and the issuing of a global ban against trading the fur of Tibetan antelopes, as part of a world-wide campaign to protect these animals.[3]

The case of anti-deforestation (snub-nosed monkey protection) is prominent and representative of this point. The anti-deforestation movement started in 1995 when domestic environmental activists and an NGO – namely Friend of Nature – drew the public's attention to deforestation and the dismal fate of snub-nosed monkeys in Deqin County, Yunnan Province, by setting up a campaign to protect these creatures in Yunnan. This campaign is thought of as having been successful, producing immediate effects: local governments changed their policies. Local governments in the Yunnan Province based their economic strategy of the "wood economy" (*mucaijingji*) on logging primary forests and selling wood. This government decision led to the disappearance of 30,000 square hectometres (hm^2) of primary forests in Deqin County (Yi 2007). In general, southwest Yunnan suffered the annual loss of 130,000hm^2 of forest because of logging (Xu, Zhang et al. 2003). At the start of this campaign in 1995, a young activist, Xi Zhinong, with the help of Friend of Nature and its founder, Liang Congjian, sent a letter to the then state council member Song Jian, appealing to him to stop the decision by the local government in Deqing County to log a thousand-acre primary forest in the region of Baima Snowy Mountain. The turning point came from the moment when Song Jian made comments on the letter and instructed local government departments to respond to the letter. The letter and Song Jian's responses were soon picked up by news media across the country for coverage. The investigative report, "Local government Started Logging Primary Forests After Having Received Huge Financial Subsidy From Central and Provincial Governments to Support Local Economy and Protect Primary Forests" (*butie daoshou futou chushou*) which was carried by CCTV's *Focus (jiaodian fangtan)* on August 2, 1998, revealed the deforestation activities in Deqing County. This report made the local government's deforestation activity well known and attracted extensive national attention. This report was the first influential investigative report on environmental problems. The campaign successfully stopped the "wood economy" policy of local governments such as the Deqin government, and prevented them from further destroying original forests in Yunnan (Liang 2012).

This story has been repeated over and over again in the coverage of news media in the years since and has been appraised as a real joint triumph of ENGOs and investigative journalism. The success is mainly credited to raising the awareness of the public and of officials at the higher administrative levels, which in turn put pressure on local governments. However, this success was a one-off and does not at all mean that local governments have completely abandoned their ambition to increase local GDP. Therefore, although deforestation was at that time discarded, other economic activities have continued and therefore pose a constant threat to original forests.

In the late 1990s, several news media outlets that started supporting investigative journalism, such as CCTV's *Oriental Horizon* (*dongfang shikong*) (1993) and *Focus* (1994) in Beijing, *Southern Weekend* and *Southern Metropolitan Daily* in Guangdong Province and *Dahe Daily* in Henan Province, paid attention to environmental problems in a random way, without dedicating particular columns or programmes to this topic. Environmental investigative reports did not differ much from investigative reports on other topics, in terms of reporting angles and writing skills, during that period of time and had a focus on limited topics, mainly pollution and deforestation.

Despite their limited number and range of topics, environmental investigative reports were often influential and effective. For example, in 1995, some factories, including the Zhutai Paper Manufacturing factory in Zhutai, Shandong Province, caused severe pollution to local residential areas in the adjacent Xingfu town. Because of the unresolved pollution, the two towns entered into fighting. The dispute continued until the publication of an investigative report by *Focus* that exposed the pollution problem in April 1996. The investigative report finally led to the problem being solved and the resolution of the conflict.[4] In 1997, the majority of news media hailed the closure of the Yangtze River and the central theme was to celebrate and praise the central government's decision to build the Three-Gorges Dam. *Southern Weekend*, however, was the first to speak out in a different, critical voice about the damage the Three Gorges Project had caused to the environment and culture with its three investigative reports "Three-Gorges Migrants Build their new lives with difficulty" (*sanxia yimin jian feicheng*), "Heritage damaged in the Three-Gorges region" (*sanxia wenwu zao pohuai xianzhuang*), "True lives of local residents in the Three Gorges region" (*sanxia de zhenshi minsheng zhuangkuang*) under a column "My beautiful Three-Gorges, my home" (*meili de shanxia wo dejia*) (Zhao 2004). This series of investigative reports is believed to have changed the state's policies, especially on relics and heritage preservation (Zhao 2004).

Entering the 21st century, in general investigative journalism has become more established than in the 1990s. Although investigative journalism on TV, such as CCTV, has seen a decline, investigative journalism in print media has witnessed the infusion of new blood, made up of metropolitan newspapers such as *Xiaoxiang Morning*, *Yunnan Information* and *Oriental Morning (dongfang zaobao)* and news magazines such as *Caijing* magazine and *New Century (xinshiji)* magazine that have become the main forces in investigative journalism (the development of investigative journalism has been discussed in detail in writings such as Tong and Sparks 2009; Tong 2011; Svensson, Saether et al. 2014; and in other Chinese-language publications such as Zhang 2010).

News media have paid more attention to environmental problems in the 21st century than in the 1990s. Environmental problems indeed have become a significant category for investigative journalism. Among influential news media regularly reporting on environmental problems are *Southern Weekend*, *Southern Metropolitan Daily*, *21st Century Business News (21shiji jingjibaodao)*, *First Financial and Economic News*, *Beijing Youth*, *Xiaoxiang Morning*, *Oriental Morning*, *Yunnan Information*, *Caijing* magazine, *New Century* magazine, *China News Weekly* magazine (*zhongguo xinwen zhoukan*), and several important TV vehicles, including *Focus* and *News Probe* (*xinwen diaocha*) on CCTV. These media outlets have set up and dedicated particular columns, pages or programmes to environmental problems. For example, *Southern Weekend* launched "Green Pages" (*lvban*) in 2009, while *Southern Metropolitan Daily* started a column called "Dark Green" (*shenlv*) in its weekly investigative report "In-depth Reporting Weekly" (*shenduzhoukan*) the same year. Relevant journalists have been allocated to undertake reports of this type. According to Qian Qiang, an experienced environmental journalist at *First Financial and Economic News*, who started reporting environmental problems in 2003, commercial news media focusing on financial and business news have become the main outlets covering environmental problems since the start of the new century. *21st Century Business News, Economic Observation (jingjiguancha bao), First Financial and Economic News* and

Caijing magazine have extensively covered environmental problems, reporting from an economic and financial angle to analyse the relationship between capital and the environment (Interview, July 11, 2011). The author's interviews with journalists and directors at media outlets such as *Southern Weekend*'s "Green Pages", *Southern Metropolitan Daily*'s "In-depth Reporting Weekly" (*shendubaodaozhoukan*) and *21st Century Business Daily*'s "Low Carbon Weekly" (*ditanzhoukan*) and *Caijing* magazine confirm that investigative reports on environmental problems or issues occupy an important position in their news organisations. The director of the environmental reporting department at a famous metropolitan newspaper, for example, said the journalists in his department were among the best investigative journalists at that newspaper and that environmental investigative reports held an important position, as reports on these topics were often published on the front pages (Interview 2011). This evidence suggests an institutionalisation of environmental investigative journalism within these news organisations.

Since 2000 it is possible to see a less influential role for ENGOs but a more active role for investigative journalism, while the relationship between ENGOs and investigative journalists became clearly symbiotic. ENGOs met their Waterloo in 2003–2004, as exemplified in the protest against the construction of dams on the Nujiang River (*nujiangfanbakangzhen*) around 2003, which is widely seen as a failure, or at least as less successful than the anti-deforestation campaign in the 1990s, and in which environmental NGOs were heavily criticised by domestic intellectuals and the news media (Sun and Zhao 2007; Zeng 2009). Since then the role of ENGOs within the environmental movement has been declining (Zeng 2009) and investigative journalism has taken over the leadership in debates and agenda-setting. There is still a close collaboration between investigative journalism and ENGOs in the 21st century. ENGOs continue to send letters of appeal to the public or to governments and officials at higher administrative levels. On many occasions, however, they act more as news sources for investigative journalists.

The efforts of environmental NGOs and investigative journalism in advancing environmental movements have not been fully repaid. The campaign against the Little Nanhai Hydroelectric Project, starting from 2009, is a classic case which environmental NGOs and investigative journalism jointly promoted the development of the movement. This project officially started in 2008. From 2009, investigative reports appeared in news media, such as "Little Nanhai Hydroelectric Project: the killer on the Yangtze River?" (*Southern Metropolitan Daily*, March 14, 2012), appealing to the authorities to rethink the development of this Little

Nanhai project and to consider protecting rare and almost extinct fish in the Yangtze Region. In 2010, six NGOs, including Friends of Nature and Green Home, formally submitted a letter to the Environmental Protection Bureau requesting to attend the Nature Protection Zones Reviewers' annual conferences at the central level that will decide whether or not finally to approve this project. Experts and academics wrote open letters, suggesting stopping the first-phase construction of the project. However, the project has not been abandoned or suspended, unlike the Nujiang project a few years ago. In 2014, the project is still under construction and the debates continue. In this sense scholars argue that the protest against the Nujiang project is not a failure at all, if compared with the Little Nanhai project campaign.

The activities of domestic ENGOs have been characterised in investigative reports more frequently than those of their international counterparts, such as Greenpeace. To take the anti-deforestation movement for example: the anti-deforestation movement has been continued by international ENGOs – especially Greenpeace, which has played a very prominent role since the beginning of the new century – while the role of domestic ENGOs that had been active and successful in maintaining this agenda in the late 1990s has declined. In 2004, Greenpeace published a report based on its investigation, accusing the Jinguang Group of having lied about its deforestation plan. This group claimed to have planned to develop barren land into 27.5 million Mu[5] of eucalyptus forests (eucalyptus is the type of tree that is needed for the group's pulp paper production). Greenpeace's report revealed that half of the land that the group claimed to be barren land was original forest and the group's plan was going to lead to logging and the destruction of the forest and species diversity. The efforts of Greenpeace soon pushed relevant government departments to respond and successfully mobilised consumers to resist products produced by the group in the following years. After nearly ten years of campaigning, in February 2013, Greenpeace forced the transnational group to promise to stop the destruction of the rainforest. However, Greenpeace's efforts have not attracted much attention by investigative journalists in China. Examining the print media coverage of Greenpeace's opposition to the Jinguang Group from 2000–2013, there have been only daily reports, and rarely investigative reports, when this international NGO has made progress.

Of course there are several reasons for the increased visibility domestic ENGOs enjoy compared to their international counterparts. First, domestic ENGOs are well connected to the news media in many ways. There is a tradition for journalists to become environmental activists and work for NGOs, as mentioned in the previous chapter. Wang Yongchen (*China's People's Radio*), Feng Yongfeng (*Guangming Daily*) and more recently Xu Nan (*Southern Weekend*) are good examples of this. Second, it is due to the agendas on which ENGOs work. For domestic ENGOs such as Green Home, to take an example, their interests are in anti-hydroelectric projects/dams in order to protect animals and rivers. This agenda is also at the heart of the current interests of environmental investigative journalists (Interview, 2013). Ning Xia, an investigative journalist at *Southern Metropolitan Daily*, for example, attended the Ten Years' Visits to Lakes and Rivers (*jiangheshinianxing*) organised by Green Home in 2013 and published a series of investigative reports on her return that revealed the destruction of nature and the miserable life the building of dams has left behind for local residents. She told the author in the interview that she was shocked by what she saw in the visit. Nevertheless, investigative journalists interviewed for this study denied that they are manipulated by ENGOs. According to them, despite having more visibility than their international counterparts, domestic ENGOs are far from being able to manipulate investigative journalism.

Topics that enjoyed prominence in the 1990s have lost their central status in the new century. This is exemplified by the opposition to deforestation, which was one of the main topics in the 1990s but is no longer prominent, compared with issues such as pollution and dam construction. The focus of the movement also changed: during the later 1990s the anti-deforestation movement was activist and NGO-oriented, whereas the involvement of news media has become more prominent in the 21st century. Newspaper coverage since 2000 on this topic, however, is characteristic of positive reporting on the achievements of protection rather than critical reports that push governments to reinforce such protection. More recently investigative journalism has turned its attention back to the topic of anti-deforestation and monkey protection when other environmental issues, such as mining, have emerged. In 2012, for example, several investigative reports were covered by *Southern Weekend, Yunnan Information* and *New Life* (*shenghuo xinbao*), revealing the great threat presented by mining activities to the existence of snub-nosed monkeys. While it is important that there are critical voices among the positive reports, these investigative reports are small in number and have not triggered true discussions at a national level.

Topics for environmental reports since 2000 are much more diverse than in the 1990s. They have expanded from mainly anti-deforestation and pollution to include a wide range of environmental hazards. These reports place an emphasis on both the fight for nature and the fight for well-being. In general, over the last 20 years these investigative reports have built up a dreadful picture of the environment at risk through constructing nine prominent agendas of environmental hazards: pollution, cancer villages, dam/hydroelectric project building, ground collapse, resource exhaustion and depletion, deforestation, desertification, water crisis and artificial land/island construction. Only after having received the attention of investigative journalists have these nine types of environmental problems been interpreted and framed as important and severe. This echoes the argument of Douglas and Wildavsky, which contends that risk selection is socially and culturally constructed (Douglas and Wildavsky 1982). One can see that all these agendas are interconnected. This is just like one of the four laws of ecology summarised in Foster's book, "everything is connected to everything else", which indicates that "ecosystems are complex and interconnected" (Foster 1999b: 118). In the construction and development of these agendas, investigative journalists have been proactive and have already overtaken ENGOs and activists to become the agenda-setters who bring these issues to the public's attention. Meanwhile, ENGOs have played insignificant roles if compared to the anti-deforestation campaigns. For most of the time ENGOs have simply acted as one of the news sources for investigative journalism. Although some of these agendas, such as the agenda of anti-dam/hydroelectric projects, were first raised by ENGOs in the 1990s, investigative reporting enables these agendas to be continued in the 21st century.

These nine agendas (see Table 2.1) have presented a big picture of environmental problems for their readers and have made the whole nation aware of existing or potential environmental risks and hazards. The construction of the nine agendas creates a discourse of environmental risk that is characterised by a fear of the destructive power of nature as a result of unresolved and pressing environmental problems and a nostalgia for the past and the homeland which have been lost under the impact of modernisation. Home becomes a perilous place, threatening people's lives, health and safety, when the environment has been destroyed and poisoned by modernisation. One's connection to the past has been cut off due to the dramatic and negative changes occurring in one's living environment that is no longer the same place as in one's memories: instead it presents a picture of dying and fragmented rivers and lakes, hollow ground, family illness and pain, the lives (of humans and animals) lost, spoiled land, disappearing forests and grassland and, by contrast, the appearance of strange artificial things that have never existed before. Therefore this discourse of risk is a discourse of warning, fear, shock, loss and nostalgia.

Table 2.1 The nine agendas in investigative reports on environmental problems since the 1990s

Agendas	Meanings
Pollution	Environmental pollution refers to the pollution of basic elements that people need for their living, including water, soil and air pollution, caused by human activities.
Cancer villages	Particular geographic areas, usually villages, in which residents, who are usually relatives or neighbours of one another, have been diagnosed with and died of cancers. News media label these areas "cancer villages".
Dam/hydroelectric project construction	Environmental problems appearing as consequences of the building of dams and hydroelectric projects, such as lakes/rivers drying up, extinction of some aquatic species, flooding, and the destruction of flora and fauna
Ground collapse	As the name suggests, ground collapse, especially resulting from over-mining of underground resources
Resource exhaustion and depletion	Over-exploitation of natural resources, especially in regions that were once known for these natural resources such as copper, coal and iron and so forth.
Deforestation	The cutting of natural forests/rain forests
Desertification	The expansion of desert areas that leads to or results from the disappearance of vegetation
Water crisis	Problems with rivers, lakes and seas, including pollution, drying up and flooding
Artificial land/island construction	The construction of land or islands along the coasts and in the seas

One interesting point arising from the examination of the nine agendas of environmental problems is that environmental problems such as climate change, ozone depletion, global warming and glacier melting are not major categories, despite being the hottest topics internationally. Interviews with investigative journalists have offered two explanations for the absence of these topics. The first refers to journalists' judgement about the interests of their readers and the environmental situation in China. Topics of this kind seem too far away from the real lives of ordinary Chinese people and thus have not been judged as the most pressing and grave environmental problems that they would care about. The second is related to the practice of investigative journalism. For domestic investigative journalism, it is not normal to investigate and report on environmental issues on the global level, especially when they are difficult to report and hard to visualise in dramatic and symbolic ways that can immediately capture the attention and compassion of local readers.

2.2 A strange paradox: the underlying political and commercial interests and logic

There is a paradox in the construction of these nine agendas. As investigative reports often expose environmental problems caused by economic activities, it might be assumed that investigative reports on such topics oppose the interests of the state and the market. However, the situation is not that simple. Investigative reports on environmental problems have presented a dual face in relation to the state and the market. They are following the instructions of the state and endorsing a new route to profits on the one hand and challenging the interests of the state and criticising capital on the other. Whereas the revelation of environmental problems such as pollution and resource exhaustion enables the state to exercise its control over individuals and institutions at all levels, criticism of the state's policies and of capital is especially prominent in reporting on agendas that reflect multiple social values in society such as the anti-dam/hydroelectric project construction agenda (this point will be discussed in detail in the next section of this chapter).

Overall, the state has not only initiated reporting on environmental problems, but also is tolerant towards challenges posed by investigative reports of this kind to its rule. The point that environmental problems are politically relatively safe topics has been confirmed in the author's interviews and by the studies of other scholars such as Wiest (Wiest 2001) and de Burgh (de Burgh and Zeng 2012). This is in part because the birth of environmental investigative journalism had a close connection with the state's desire to clean up the environment. This is also because most environmental investigative reports criticise individuals, businesses and governments at the local level most of the time, target individual economic activities and seldom touch on political issues that are central to the state (Interviews, 2011–2013).

A close examination of the nine agendas constructed by investigative reports over the past 20 years reveals an overall match between the central government's environmental policies and the agendas constructed by investigative reports in both the 1990s and the 21st century. Therefore these agendas have a certain level of political legitimacy and march in step with the central themes of propaganda. But of course this point is more characteristic of some agendas, such as anti-deforestation and pollution, than others, such as anti-dam construction, and more relevant at some times than others.

Take the anti-deforestation case- already discussed in the previous chapter for example. If we look into the background of CCTV's 1998 report on the deforestation in Deqin County, we find that there was a natural disaster – extremely severe flooding – from June to August that year. A research report published by the Ministry of Water Resources in 1999 pointed out the connection between deforestation in the upstream area of the Yangtze River and flooding in the major rivers, especially the Yangtze River. Some 60%–70% of economic activity in Dege County beside the Jingsha River, upstream of the Yangtze River, came from the "wood economy" (wood logged from forests). In 1999 alone, the state-owned forest and industrial enterprises logged 300,000 square meters of wood, which was equal to 50,000 Mu of original forests. This pushed the central government to think about the environmental problems caused by deforestation and in 1999 it commanded all 51 major forest and industrial enterprises to stop logging, and plant trees instead (The Ministry of Water Resources of the People's Republic of China 1999). There was then a propaganda request from the top that the news media across China should extensively report the policies of "Restoring Farmland to Forest" (*tuigeng hailing*),[6] "Restoring Pasture to Grassland" (*tuimu haicao*) and the prohibition against logging original forests (*tianran lin*). This is because the state also expects investigative journalism to take on the role of public opinion supervision, which has been a tradition of the CCP, in order to maintain the "purity" of the Party and to correct the wrongdoings of Party members. Although for most of the time only positive reports such as those depicting "role models" who have done a good job in restoring farmland to forest are seen in the news media, as requested by the China Forestry Bureau,[7] such policy and natural disaster backgrounds give justification to more investigative reports on deforestation and desertification. What happened in the context, for example, the natural disaster of floods and the central government's policies, also guaranteed the success of the reports and the campaign.

In addition, investigative reports on environmental problems also match the needs of the state in highlighting the importance of what happens to the environment. Year in and year out over the past decades, the more prominent the appearance of environmental problems and the occurrence of natural disasters, the greater, more pressing and indispensable the need for a clean and safe environment has become. While being premised with the need to sustain and increase economic growth, the central leadership also wishes to reduce the harm economic activities are causing to the environment. We can see this wish being expressed in many ways, ranging from the launch of the environmental protection centennial tour, the establishment of the Environmental Protection Ministry (upgraded from the Environmental Protection Bureau in 2008) and the issuing of a regulation on Publishing Environmental

Protection Information (*huanbao xinxi gongkai banfa*) in 2007 that has driven the exposure of environmental information. The construction of the discourse of environmental risks in investigative reporting identifies and defines what can be done but also what cannot be done for the sake of the environment. On the one hand, the formation of knowledge of this kind educates and sets standards for individuals and institutions in society about their activities. This is a process of binary division and knowledge formation in Foucauldian terms about what can and cannot be done. On the other hand, the discourse of environmental risks also reminds individuals and institutions of the existence of monitoring and surveillance, here embodied in the public opinion supervision function of investigative journalism. The work of investigative journalists in exposing environmental problems caused by human activity demonstrates that the state has eyes and ears that observe and monitor society. In this sense, following Foucault's understanding of discourse and power, the discourse of environmental risks helps the state tighten its control through forcing individuals and institutions at all levels to discipline themselves according to the standards set in these reports. This, to a great extent, contributes to the consolidation of the rule of the state.

However, of course not all agendas function in this way. Investigative reports on some topics, such as pollution and resource exhaustion are closer to the interests of the state than in others, such as the agendas of anti-dam and anti-artificial islands construction. This is partly because activities like dam and artificial islands construction mesh with the policies and strategies of the state and are encouraged by the state. For example, since the publication of the Development Plan of National Marine Economy (*quanguo haiyang jingji fazhan guihua gangyao*) in 2003, the state has encouraged the exploitation of marine resources for advancing economic growth, and one primary example of this is constructing artificial land and islands that create space for making more profits. Criticising the construction of dams and artificial islands directly therefore challenges the interests of the state and of the market. However, in terms of pollution, both the central government and local governments welcome the exposure of pollution problems. For local governments, exposing pollution in news media coverage might help them achieve more financial support from governments at higher administrative levels, while for the central government, most pollution problems occur at local levels and the presence of pollution in news media coverage rings a warning bell for local governments, reminding them of environmental protection and showing the determination of the central government to fight pollution, which is good for the legitimacy of the CCP (Interviews, 2011–2013).

Even for capitalists, although investigative journalism agendas such as pollution and resource exhaustion may impair the interests of some economic groups, they offer an opportunity for other groups to gain profits. In terms of the interests of the market, revealing environmental problems caused by human activities is often associated with the criticism of capital – especially if investigative reports mainly attribute environmental problems to individuals' and governments' crazy pursuit of profits. Therefore it is true that when investigative reports expose environmental problems, most of the time the reports criticise the unsustainable economic activities of capital in ruining nature. However, there is another side of the coin. The drive of capitalism to maximise profits results in the emergence of new markets that are based on the assumption that the environment needs to be clean under current circumstances. That is to say, the picture of the spoiled environment painted by investigative reports has a tendency to create new markets for certain products and services, such as clean technologies and energy. For example, with the increasingly worsening air quality, air-cleaning equipment has great potential in the market. The sale of air-cleaning equipment increased from RMB 220 billion in 2011 to RMB 530 billion in 2013.[8] This could produce the ironic situation where the more machines of this kind are produced, the more pollution the manufacturing process will cause to the environment, driving ever more demand for the equipment. Thus humans could be seen to be facing a vicious circle in which we are creating our own doom. This circle demonstrates the continuing confidence in technologies, and the logic of relying on such technologies for human survival, with the aim of meeting the needs of humanity instead of saving the dying environment. This logic is the logic of capitalist production as well as of consumer culture and of the lifestyles of the middle and upper classes in Chinese society. This logic also explains why environmental investigative reporting has achieved much more autonomy and tolerance than other types of investigative reporting.

How investigative reports on environmental problems mesh with the interests of the state and the market can be best exemplified in the construction of the two agendas of pollution and resource exhaustion. These two agendas of investigative reporting not only match the environmental policies of the state and the strategies of environmental governance, but also make use of the reporting opportunities created by the occurrence of environmental incidents or of natural disasters. In addition, these two types of environmental problems may even lead to the rise of new markets, which fits the need of capitalism for profits. The topic of pollution (including air pollution [haze/PM2.5], soil pollution and water pollution) has been one of the main topics for environmental investigative journalism and for the state's environmental governance since the 1990s. This agenda remains consistent with the central reporting theme of revealing and reducing pollution from the 1990s onwards. Air pollution was not a prominent agenda item for investigative journalism at the beginning of the 21st century. Most of the investigative reports mentioning air pollution at that time were merely about general environmental problems faced by some geographical areas, of which air pollution was merely one type. The agenda of air pollution did not fully develop until the issue of smog and later PM2.5 emerged prominently in the public domain in recent years.

At the beginning of the 21st century, most of the reports on smog or haze were of the hard climate news type, with the aim of alerting residents about the weather conditions. Investigative reports on smog started appearing from 2006 in *Southern Metropolitan Daily*. Since then this newspaper has paid more attention to bad air quality than other newspapers with similar market and editorial stance. This is partly because Guangdong is one of the most severely affected areas and partly because of the critical tradition of this newspaper. From 2006–2013, almost every year *Southern Metropolitan Daily* has published at least one in-depth and lengthy report to investigate and interpret the causes, severity and consequences of haze in Guangdong. From 2011, when the issue of PM2.5 particles suddenly became particularly prominent on Weibo and online public opinion pushed the central government to publish data on air quality, other newspapers that favour investigative reporting such as *Southern Weekend*, *Beijing News* and *21st Century Business News*, have also published investigative reports revealing the major smog problem in big cities such as Beijing, Guangzhou, Shenzhen and Shenyang and appealing to people to stop using equipment such as private boilers or cease some activities such as straw-burning.

If compared with air pollution, a considerable number of investigative reports on water pollution started appearing from 2000, and especially from 2006. Investigative reports on this topic have moved beyond daily news reports on water pollution incidents to carry out investigations and reveal the overall situation of water pollution. As well as investigative reports triggered by pollution incidents such as the Tuojiang, Chifeng and Yancheng pollution incidents ("Investigating Water Pollution in Yancheng", *Beijing News*, February 2, 2009; "A Crisis Caused by Heavy

Rain? Investigating Water Pollution in Chifeng", *Southern Metropolitan Daily*, August 5, 2009), reports have reminded local residents of the poor situation of rivers and lakes and underground water from year to year. Most investigative reports are on local water pollution. To take the case of pollution in Dianchi, in Yunnan, for example, investigative reports on this have been covered mostly in *Yunnan Information*, which kept publishing investigative reports on this subject in 2008. At least seven important investigative reports on the topic were covered in 2008, depicting the big problems Dian Chi and local residents were facing. It is interesting to know the reasons why the local newspaper felt able to report the biggest environmental issue in that locale. According to the director of a news reporting team at the newspaper, the local government and relevant officials wanted to have the environmental problems exposed in media coverage so that they could apply for subsidies from the top in order to deal with the issue. This is to say, local officials need investigative reports of this kind. Nevertheless, the central government also desires media supervision on local officials with regard to environmental problems and issues. Environmental investigative journalism has made good use of the differentiated needs of the local and central governments. In 2012, therefore, both *Southern Metropolitan Daily* (the parent newspaper of *Yunnan Information*) and *Yunnan Information* shifted their attention from the pollution itself to environmental governance and associated corruption. The reports, such as "Vanity Fair of Dianchi: water pollution remains despite over 1,000 billion investment over the past 20 years, now it requires public funding" (*Southern Metropolitan Daily*, August 16, 2012), "Dianchi: start paying back the debts" (*Yunnan Information*, November 15, 2012) strongly criticise the local government in Yunnan for failing to deal with the pollution in spite of having claimed huge public subsidies.

Nationally influential newspapers such as *Beijing Youth, Southern Weekend* and *Southern Metropolitan Daily* have also made national readers aware of the water pollution situation in other regions. They take advantage of the cross-regional media supervision role of investigative journalism. As a result, important investigative reports for the first time are able to present an overall picture of the problems with China's great rivers and lakes. For the Taihu Lake, to take an example, interestingly, while most daily reports took a positive view of its water quality, some investigative reports did raise critical voices. The only critical investigative reports we can find from 2000–2013 occurred in 2007, triggered by the high level of Cyanobacteria the Taihu Lake was experiencing that year. *Oriental Morning* and *Southern Metropolitan Daily* covered in-depth investigative reports on this issue, making it nationally well known, though local media kept silent. Another example is the Dongtinghu Lake. Although having not reported extensively on pollution in Dongtinghu, investigative journalists from out-of-town newspapers such as *Xiaoxiang Morning* and *Southern Metropolitan Daily* revealed other severe and equally, if not more, important environmental problems such as the "mouse disaster" (too many mice appearing in the areas, destroying crops, plants and even anti-flood levees [*shuhuan*] and drought which have been brought about by the construction of the Three-Gorges Project.

People can sense air pollution and water pollution from their direct experience. However, soil pollution is invisible. Although occasionally there are daily reports on soil pollution, and policies have existed on soil pollution since 2000, the surge in investigative reports on this topic from 2010 formed this agenda in public discourse. These reports have drawn a dramatic picture of the causes and especially consequences of soil pollution that people would otherwise not know about. Heavy-metal pollution in soil has been brought into the centre of public attention in the last few years, with reports showing that not only vegetables, rice and fruit but also whole residential districts have been contaminated by industrial activities. One after another, these investigative reports have reminded ordinary people of the problem of invisible soil pollution and the association between pollution and diseases such as cancers, as well as the hidden truth that some real estate developers have constructed buildings on polluted soil and that some farmers grow vegetables and crops on contaminated soil. From 2010 onwards, polluted land has been renamed "toxic land" (*dudi*) and has been revealed to have been used for residential buildings without telling buyers. Reports such as "Can thousands of buildings be safely built on poisonous lands? (*Southern Weekend*, December 12, 2011) and "Wuhan Estate Company Built Houses on Heavily Polluted Land, All houses sold out but buyers unaware of the danger of poisonous lands" (*Life Daily*, December 1, 2010) have revealed this problem. In terms of contaminated food, influential investigative reports, especially "Risky Cadmium Rice" by Gong Jing at *New Century* magazine in 2011, first raised public concern over food security and attracted the national news media's attention to this issue. Investigative reports on this topic of pollution are also often associated with the topics of cancer villages or other health problems such as high levels of lead poisoning that may be caused by environmental problems. Media attention to this issue reached a peak during the period February–May 2013. Taken together such reports depict a picture of pollution that has raised the awareness of the general public about these issues. Therefore it becomes a pressing question how to clean air, water and soil.

As far as resource exhaustion is concerned, the state started paying attention to this problem in 2007. Meanwhile, investigative journalists also started reporting on this topic. The issuing of Guidance on Promoting the Sustainable Development of Cities Relying on Natural Resources for Economic Development (*guanyu chujin ziyuanxing chengshi kechixu fazhan de ruogan yijian*) in 2007 signals that this economic agenda had become central to China's economic development and environmental protection plans.[9] According to this guidance, those cities that have been defined as "resource exhaustion cities" (*ziyuan kujie xing chengshi*) will receive financial support and reimbursement from the state. This implies that the disclosure of resource exhaustion problems in locales in fact can help local governments obtain funding from above. In 2013, it was reported that the central government had offered RMB 1,680 billion to support the transformation of resource-exhausted cities, that is, the rebuilding of other types of economies in these cities. This very detailed dynamic explains why investigative journalists are able to report on such a problem. This echoes with an interesting common point raised in interviews that the disclosure of local environmental problems on some occasions benefits rather than threatens local governments because they can receive subsidies from the government at higher administrative levels. Ironically, a news report covered by *Ningxia Daily* claimed: "after more than three years' continuous hard efforts, Shizuishan City, that depended on coal resources for its development, has been finally successfully ranked as one of the first 12 resource exhaustion cities" ("Shizhuishan was ranked in the first list of resource-exhausted cities", *Ningxia Daily*, March 27, 2008). The triumphant tone embodied in the report reflects the local government's desire to achieve the financial support of the central government by being designated a resource-exhausted city. The first list of 12 resource-exhaustion cities was published on March 20, 2008, while the second list of 12 resource-exhaustion cities was published on March 5, 2009. Soon after the publication of the lists, investigative journalists intensified their reporting on this topic. Famous reports include "Copper Capital Daye: an exhausted treasure bowl" (*Beijing News*, March 18, 2009); "Buxin after 'the coal sea' exhausted" (*Beijing News*, March 20, 2009); "The 20-year struggle of Dongchuan after three cases of copper mine exhaustion (*Yunnan Information*, March 30, 2009); "Sand stealers took treasures from Dongjiang: mother river is facing sand exhaustion, saltwater flows in, difficult to get river water along the bank" (*Southern Metropolitan Daily*, September 15, 2009); "The breakthrough of resource-exhausted Leiyang" (*Xiaoxiang Morning*, April 23, 2009); "Dilemma of 'Black Gold': incense becomes popular, overseas stealers crave Primary Forests in Xishuangbanna, leading to resource exhaustion" (*Yunnan Information*, November 16, 2010); "Baining after resource exhaustion" (*Yunnan Information*, February 26, 2013); "Water exhaustion caused by coal mining: almost ten thousand villagers depend on rain water for living" (*Dahe Daily*, August 16, 2013).

Apart from chiming with the environmental policies of the state, these two agendas are also related to the rise of new markets that can open new doors for capitalism. Solving pollution, to take an example, requires clean technologies and products. Polluted water, soil and air call for human efforts and innovation to solve the problems. This implies confidence in human ability to deal with problems with the environment, which is the logic that the state constructs for ecological modernisation and that the market shapes for exploring new markets and maximising profits. These new markets also generate appetites for new energy sources to replace those traditional ones such as coal, wood and oil and for new resources to replace those such as copper and iron that have run out.

Therefore, there are political and even economic justifications for those environmental agendas constructed by investigative reporting. For investigative journalists and news media, environmental problems are an excellent topic to report on – they reflect all kinds of problems in Chinese society and meet the criteria of telling good stories. This therefore not only fits their ideals of professional practice, but also meets the requirements of the media market by satisfying demand from ordinary readers. Moreover, according to investigative journalists interviewed for this study, such reports will not affect existing advertising, as enterprises also want to create a "green image". Reports of this kind can even attract new advertisers – especially those working on clean technologies and new energy sources. This point shows again that these investigative reports can also be in the interests of the market. However, the meshing of this type of journalism with existing political and commercial interests does not negate their warning function. Although investigative reports on pollution (air, water and soil pollution) may fit the interests of the state and open new markets for new businesses such as clean food and new technologies that can clean the air, the water and the soil, the portrayal of pollution in investigative reports overall inevitably paints a picture of environmental risks that may scare the public.

2.3 Environmental controversies, critical voices and investigative journalism

Investigative reports on environmental problems do not always accord with the interests of the state and the market. They can oppose these political and economic interests in part because they mediate controversial debates on environmental issues, central to which is a reflection on a basic philosophy in Chinese society surrounding whether humans should conquer or respect nature, that is, how Chinese people should position themselves to nature. This is a reflection on the classic Promethean discourse promoting rational and sustainable development. Such a reflection is underpinned by multiple interests that care more about the well-being of humans, by popular worries of risks caused by the construction of large projects, such as dams and artificial islands and lands, and by environmental scares and fears for consequences such as those seen in cancer villages. A call for a rethinking of the human–environment relationship underlies such a reflection.

Thus, in constructing some environmental agendas, investigative journalism is driven by diverse voices and values that seek an alternative human–environment relationship and that reflect the deep fear of environmental catastrophes. Prominent agendas of this kind include those which oppose dam/hydroelectric projects and artificial islands. Compared with the environmental movement of anti-deforestation and animal protection, investigative journalism is much more proactive in advancing the case against the construction of dams/hydroelectric projects and has given this agenda a great deal of attention and coverage, especially since 2000, and reports on these topics voice strong criticisms towards the central and local governments' policies. The critical role of investigative journalism is shown in two ways: first, investigative reports have mediated diverse, even opposition, voices regarding the construction of dams, even while there has been a nationwide frenzy of dam building. Investigative reports have incorporated and presented multiple voices from a variety of news sources, rather than expressing their own viewpoints against or in favour of dam projects. In this sense, environmental NGOs, activists and scientists can be heard when they are allowed to enter media coverage, while investigative reports can thereby obtain their authority for recognizing the associated environmental problems. Second, investigative reports raise critical voices regarding the building of dams and/or the awareness of the risks that certain government policies may give rise to when the majority of the public is unaware of the hidden truths that relevant authorities want to cover up. The critical voices that are embodied in these two agendas truly diverge from the mainstream values promoted by the state and the market. The critical voices are part of diverse voices in Chinese society that not only come from the bottom of society, but from the leadership and elites. The rulers themselves are not a monolithic group; they have different views and voices. Environmental investigative reports have become one of the main sites where critical voices are heard.

The anti-dam movement can be traced back to the late 1990s and the turn of the 21st century; critical voices opposing the construction of dams and hydroelectric projects have been heard all the time. From the Yangliu Lake hydroelectric project at Dujiang Yan, hydroelectric developments on the Nujiang River, to the Three Gorges Dam, investigative journalism has played a proactive role in advocating for the anti-dam movement. Examining investigative reports on environmental problems over the past decade, opposition to dam building is the number one agenda, to which investigative journalism has made a great contribution. The contribution of investigative journalism lies in airing critical and questioning voices when the construction of dams has been legitimised by the decisions of the central government. In addition, investigative reports have also drawn the attention of the public to the disadvantages of constructing dams, the hidden power abuses involved as well as the exploitation of disadvantaged social groups concerned.

The anti-dam movement exemplifies very well the conflict between "being rich"/"economic development" and "environmental protection". Intensive debates surrounding the construction of dams, especially the Three Gorges Dam, have not stopped since the 1950s. The fierce advocacy of the construction of dams reflects a prominent environmental trait of socialism (also collectivism) in China. The anti-dam movement for its part originates from the historically rooted controversial debates over the construction of the Three Gorges Dam.

The Three Gorges Dam is the biggest dam that has ever been built in the world. Mao's poem: "hold back Wushan clouds and rain when lake rises in the high dam" (*wushan jieduan yunyu, gaoxia chu pinghu*), showing praise and approval for the construction of the Three Gorges Dam in 1956, reflects his eagerness to conquer nature for the sake of development and his belief in humans' capacity to do so. His enthusiasm for constructing the Three Gorges Dam, however, was deflected by opponents, such as Li Rui, in debates in the 1950s. Since then there have been two more major debates, in the 1980s and in the 21st century. In the 1980s there was a certain amount of room for free speech as a result of the prevailing liberalism and the start of commercialisation and the opening-up reforms before 1989. International NGOs such as the International Rivers Network (IRN) also played an active role in advancing the anti-dam movement at that time (Wang 2005). However, the decision to go ahead with constructing the Three Gorges Dam was actually made at that time and thus the 1980s' debates were thought of as crucial (Tang 2009).[10] The 21st century, especially since 2002, has witnessed continuous debates which have questioned the construction of a number of hydroelectric projects. Investigative journalism has kept the debate going, rather than letting the issue be wholly controlled by the side that supports the construction of dams.

Although the building of dams chimes with the economic development strategies of the central government, a number of news media that are well known for their investigative journalism practices have made space available for critical voices opposing dam construction, as well as exposing wrongdoings involved in the process of dam construction. The anti-Yangliuhu hydroelectric project in 2002 signals the start of the movement in this century. The Yangliuhu hydroelectric project was a crucial part of the Zipingpu hydro project, a symbolic project in the state's Western development plan (*xibu dakaifa*). A close examination of newspaper coverage of the Zipingpu hydro project (including the Yangliuhu project) from 2000–2013 reveals that there were hardly any critical voices against the construction of the Zipingpu project when it was approved and started in 2000. Three years later, however, a piece of news in *China Youth* (*zhongguo qingnian bao*) by Zhang Kejia titled "New dam will be constructed, the original site of World Heritage Dujianyan Dam will be destroyed, UN gives attention" (*shijie yichan dujiangyan jiangjian xinba yuanmao zaopohuai lianheguo guanzhu*) on July 6, 2003 revealed the possible damage the planned new dam associated with the Yangliuhu hydroelectric project may cause to Dujianyan Dam and thereafter sparked a national wave of doubt over the construction. Nationwide news media sent their journalists to report on this issue. Prominent among them, *Southern Weekend*'s investigative report further exposed and criticised relevant government departments and officials' insistence on constructing the Yangliuhu Dam regardless of the opposition of experts ("Dam is only 1,310 meters away from Dujiangyan", August 1, 2003). This campaign effectively led to the suspension of the local government project. This was regarded as a triumph of the news media ("Who Protects Dujiangyan?" *Southern Metropolitan Daily*, September 22, 003). These reports embody the wider public disagreement over the construction of the dam. The Yangliuhu project, nevertheless, has disappeared from newspaper coverage since it was mentioned in *Southern Weekend*'s investigative report: "Investigation into the violation of Jinshajiang Hydroelectric Station" (October 29, 2010) which revealed the wrongdoings of the electrical companies. The environmental protection ministry already knew about these wrongdoings and had commanded them to cease, which means this report principally met the requirements of the central government, reflecting central–local government struggles.

Similarly, in the next year, 2004, local drives towards "being rich"/"economic development" came into severe conflict with "environmental protection". Investigative reports carried by critical news media such as *Southern Weekend* ("Hutiao Strait in danger: one of the most beautiful natural scenes may disappear because of the construction of dams", September 30, 2004), *Southern Metropolitan Daily* ("Nujiang will also have new dams", November 25, 2003) and *Beijing Youth* ("large hydroelectric projects left 10 of 16 million migrants in poverty", July 29, 2004) cast doubts on hydroelectric projects on the Nujiang River, expressing controversial debates among experts. The critical voices embodied in these reports in critical news media strongly contrast with positive news reports on the same issue covered by party organs. In fact, examining newspaper coverage on the Nujiang Dam from 2000 to the present time reveals distinct discourses among party organs and investigative reports. These contrasting views suggest China has multiple values and voices rather being monopolised by one value and voice that the political authorities approve.

The second time in the new century when extensive critical voices about the construction of dams were again heard was during and after earthquakes such as the Wenchuan earthquake. A representative investigative report of this kind is the one published by *Southern Metropolitan Daily*: "Controversies over the reconstruction of hydroelectric projects in the Longmenshan Fracture Zone: opponents required reassessment of the safety of construction, while supporters believe the Wenchuan earthquake reinforced confidence in constructing hydroelectric projects in high mountain straits" (June 27, 2008). This report also concerns the impact of the Three Gorges Dam on the geology. Although there were disapproving voices raised against the construction of Three Gorges in the 1980s, unlike other dams the Three Gorges did not receive much critical coverage about the dam itself in the period since 2000 (most critical coverage focused on social problems such as the impact on migrants and corruption problems). From 2009 to 2010, newspapers across China reported an official evaluation report that claimed there is no necessary relationship between the Three Gorges Dam and earthquakes. By contrast, since 2010, after several earthquakes and other natural disasters such as drought have struck China, investigative reporting has started linking the giant dam to these disasters and questioning the legitimacy of the dam itself. In doing so these investigative reports directly questioned the central government's decisions to construct giant dams and tried to stop governments from making further mistakes. Representative reports among others include "Risky situations in Three Gorges" (*Southern Metropolitan Daily,* March 3, 2010), "Debates over Three Gorges drought" and "Abnormal Three Gorges" (*First Financial and Economic News,* May 31, 2011), "Risks facing China's giant dams?" (*Southern Weekend,* July 7, 2011), "Three Gorges' next stop" (*Southern Metropolitan Daily,* June 8, 2011), "Investigation into flooding and landslide in Jinshajiang Baihetan hydroelectric project construction site: no warning received" (*Xiaoxiang Morning,* July 6, 2012) and "Ten years too short to decide what Dongtinghu Lake needs" (*Xiaoxiang Morning,* April 2, 2013). Meanwhile, certain Party organs started responding to the increasing doubts about the Three Gorges, trying to distance dams from natural disasters and attacking these "rumours", for example, "Three Gorges has brought benefits such as flood protection and electricity generation" (*Global Times,* September 2, 2013). In 2011, China for the first time officially admitted the association between disasters and the construction of the Three Gorges and other dams.

The intensive investigations into the building of dams/hydroelectric projects reflect different views among high-ranking leaders, intellectuals, scientists and the public in the wider social context. The neo-leftists' *Xinyusi* website[11] has become the main forum, holding debates among intellectuals and representing different interests on the construction of dams. Investigative reports have therefore become a major site in which different views on the relationship between development and nature are reflected. This perspective reflects the multiple cultural values prevailing in Chinese society. In addition, it also raises concerns over the exploitation of disadvantaged people and the preservation of cultural relics and heritage. However, these reports do not fit with the mainstream policy of the central government that has been in favour of hydroelectric development and dam building. Given this background, investigative reports on this topic seldom have the effect of resulting in changes to policies. Even if there are changes in local governments' policies, the effects are usually one-off and temporary, as exemplified in the anti-Yangliuhu project movement.

Compared to the voices raised against dam building, public opposition to the construction of artificial land and islands is relatively weak. Nevertheless it became stronger after 2010, expressed not only by critical news media such as *Southern Weekend* (such as "Great leap of sea reclamation", August 11, 2011) and *Southern Metropolitan Daily* (Sea reclamation: how to stop hurting mangroves, September 3, 2008) but also by commercial newspapers such as *Evening News Today* (*jin wanbao*) ("Sea reclamation: who is crazy for buildings built on artificial lands", April 9, 2010). Despite being relatively weak, the critical and oppositional voices have been spelled out by investigative reports after all.

2.4   Environmental justice, scares and labelling: the agenda of "cancer villages"

Seven members of a family died of cancer in Shimen County, Hunan Province. In five of the cases, their cancers were proved to be caused by arsenic in the environment. More than 1,200 among some 3,000 local residents in the county have been found to have been poisoned by arsenic. The main culprit is realgar ore (an arsenic sulphide mineral) from a local sulphur plant. From 1971 to January 2013, more than 400 staff members working in the local realgar ore mine died of cancer. The former prosperity in the place where the plant and mine were located has long gone and only left a poisoned environment and the notorious name of "cancer village" for local residents. This is the miserable scene described in a recent report in *Law Evening News (fazhi wanbao)* (April 13, 2014). This is also the typical scene that most investigative reports on "cancer villages" depict, including a destroyed environment, severe health hazards, scares about incurable diseases, a group of local residents who cannot leave and are left there to rot and die, the lack of environmental and social justice, and the label of "cancer village".

Emerging from nowhere, the topic of "cancer villages" has gradually become prominent on the public agenda since 1998. Unlike other agendas, such as the construction of hydroelectric projects/dams, this issue of "cancer villages" is not exclusive to critical investigative reporting; instead it has also been covered both by Party organs and by critical commercial news media over the past decade. This is the only agenda concerning environmental problems to which both party organs and investigative journalists have given extensive attention. "Cancer villages" is a term news media use to label geographical areas where a large group of local residents have collectively been diagnosed with cancer. Usually such a geographical area is a village and those who suffer from the cancers are villagers. Therefore the term implies that the victims are often those who live in remote areas rather than in urban cities.

From 2000 to 2012, press coverage on the topic of cancer villages steadily increased, and there was a sudden surge in the number of reports in 2013. The number of news reports on cancer villages in 2013 is nearly equivalent to the total number between 2000–2012. Fewer than ten investigative reports on this topic were published each year before

2010 and, in some years, only one or two. The number of investigative reports on the topic reached 13 in 2010. Both 2011 and 2013 saw the highest numbers (more than 40) of investigative reports on cancer villages due to an emerging pollution incident in 2011 and the official confirmation of the existence of such cancer villages and of the connection with pollution in 2013. In 2013, the number of general news reports on cancer villages suddenly increased. At the time of this writing (November 2013), 1,221 news articles mentioned this topic, while 209 reports appeared in 2013 (there were only 135 articles in 2009–2010, 51 in 2004–2008, 38 in 2000–2003, with "cancer village/s" appearing in the titles). This surge in the number of reports is credited to the publication entitled "Water Environment Along Huai River and Death Map of People Who Got Gastrointestinal Cancer in the Region" (*huahe liuyu shui huanjing yu xiaohuadao zhongliu siwang tuji*) and the acknowledgement of the existence of cancer villages by the Health Ministry in 2013.

Investigative reports on "cancer villages" in the early years of this century are usually based on journalists' own investigations, on-site scenes witnessed by journalists, and information provided by victims, such as lists of the dead, with the aim of revealing the horrible reality to the external world. These reports include "Shangba Village: the source of life is dead" (*Yangcheng Evening*, March 10, 2001), "Who made this village a cancer village?" (*Science Daily*, May 19, 2004), "Investigating cancer villages: severe air pollution, people need to cover mouths and noses in sleep" (*Beijing Youth*, April 26, 2004). These investigative reports directly attributed the appearance of cancer villages to industrial pollution.

Over more recent years investigative journalists have begun to bring in scientists' research studies to complement their observations on those "cancer villages". Prominent examples of this are the reports by CCTV's *Economic Half Hour* ("Hengshi River flows across village of death," June 2, 2005) and *Southern Metropolitan Daily* ("Help and hope of cancer village", November 18, 2005; "Cancer village beside Hengshi River: an unrecoverable wound of the Zhu River", July 22, 2009) on Shangba Village, Guangdong Province. These reports established a connection between mining and the high occurrence of cancers by citing a scientist's study that examined the quality of the Hengshi River and concluded that the river water was toxic.

In these ways investigative reports convey an intense critical voice that reflects on the negative consequence of practicing capitalism in China, raise the general public's consciousness of the danger posed by economic activities to people's heath and call for more care for Chinese people's well-being. The miserable life and fate of cancer victims is presented vividly. Through constructing agendas such as "cancer villages", investigative reports are also seeking environmental justice for victims in "cancer villages". According to journalists interviewed in the study, these victims are usually from the bottom of society. They are illiterate, poor and unable to leave the environment that has caused damage to their health. They are not even capable of seeking environmental justice for themselves. Thus it is to investigative journalists that they turn for help.

However, repeated reports of the same problems in the same places suggest no that effective action has been taken since the issue was first raised. Investigative journalists at some news media, such as *Southern Metropolitan Daily*, have visited villagers living in cancer villages several years after their previous interviews to see whether or not the situation has changed. The repeated visits and reports show the ineffectiveness of these reports. *Yangcheng Evening* first revealed the problems with the health of villagers in Shangba Village and the threat posed by the environment in 2001, and after 13 years, in 2013, investigative journalists are still reporting on the cancer problem in the village, such as in the report "Dilemma in cleaning polluted soil in Northern Guangdong" by *Beijing Youth* (August 29, 2013). These investigative reports, most of the time, link the occurrence of cancer villages to pollution – especially water pollution and soil pollution – caused by China's industrialisation activities.

The potential for a change in this situation came in 2013 when the central government for the first time admitted the existence of cancer villages. It is the news media that placed the agenda of cancer villages into the public arena. However, the agenda had not influenced political decision-making until it became central to online attention. Over the past few years, especially, unofficial cancer village lists and maps have appeared on the Internet. In 2013, the map of cancer villages became even more popular. It was in 2013 when online discussions on cancer villages, especially the map of cancer villages, entered investigative reports for the first time. Before 2013, in 2011 and 2012, although news reports sometimes associated cancer villages with online discussions, they often labelled such online discussions rumours. Online discussions and information on cancer villages started to be treated as credible news sources in news reports from 2013.

However, the label of "cancer village" implies for the most part that it is people who live in rural areas that suffer pollution-related cancers. Therefore, although investigative reports on cancer villages are very striking, for most urban residents, aside from their humanitarian sympathy for cancer victims, the problems about cancer villages basically do not seem have anything to do with them, as their living environment is far away from those rural areas. This thus alienates cancer victims from urban residents, especially the middle and upper classes. This alienation might be able to stabilise society if the middle and upper classes are seen as the pillar of Chinese society. However, if the lower classes are the most important foundation for social stability, such an alienation may impair the basis for the rule's legitimacy.

2.5 A complex picture: a self-contradictory role of investigative journalism

China is like a ship sailing in the sea. This ship is too big to have all people working at the same pace and to turn the ship in time to avoid damage. Environmental problems represent damage occurring on the sailing journey. Environmental investigative journalism is one of the effective tools the ship's captain (the state) uses to explain what is happening on the journey (modernisation process) to the passengers (Chinese people) and make them understand what to do.

Over 20 years of development, we find the emergence of environmental investigative journalism results from the joint forces of the state, the market and society. In Marx's view, nature is a "free gift" to capital, which suggests it is in the nature of capital to exploit nature for profits. Foster thus sees a natural rift between capital and nature. In fact, the Chinese central government has realised that there is such a rift between capital and nature. From the development of environmental investigative journalism in China, we can see the important role the state has played in prohibiting the exploitation and damage capital has done to nature, which means the state and capital cannot be truly conspiring against the environment. However, such a role is limited by the state's desire for modernisation. Environmental investigative journalism has developed in this context. Media marketisation also draws the attention of investigative journalists to environmental problems that are central to the interests of ordinary viewers and readers.

The development of environmental investigative journalism over the past 20 years presents a self-contradictory picture regarding the role of journalism in portraying the rift between modernisation and the environment. The relationship between investigative journalism and the state is the starting point for critical analysis of the role of environmental investigative journalism and offers a prism through which we can seek answers to this question. A traditional view sees Chinese journalism acting as a state ideological apparatus that constructs and diffuses the dominant ideology and sustains the hegemony and rule of the state. The realities of media control in China ostensibly prove this argument to be true. At some point these environmental investigative reports somewhat play the role of a state ideological apparatus. The overall coherence between environmental policies and environmental investigative reports reflects the ideological harmony between environmental investigative journalism and the state. The public opinion supervision function of investigative journalism exercises the state's control over individuals and institutions by making them self-discipline their activities. The association of environmental problems with rural areas alienates environmental victims from urban residents, which may or may not contribute to social stability. The revealing of environmental problems by investigative reports somewhat justifies and nurtures new markets for new clean technologies, products and energy.

Nevertheless, if this account is absolutely accurate there is no way for Chinese investigative journalism to function as a risk informer that reports the situation of environmental problems to ordinary people and even interprets the meaning of environmental problems for them. However, the fact is that investigative reports have already seized the chance and constructed agendas of environmental problems over the past few decades. The agendas contribute to the forming of the big picture of environmental risks, helping people in the lower classes and dissidents to make their voices heard. Behind the critical investigative reports on environmental problems is the prevalence of plural social thoughts and of different views of the human–nature relationship. Investigative journalism has seized this chance. The environmental crisis of China, as constructed in 20 years of investigative reports, constitutes a strong message that critical voices against inappropriate and unsustainable economic activities/growth are tolerated by the state. Investigative reports on environmental victims give them emotional support with the aim of seeking solutions for them. This requires us to understand why there are complex and even contradictory situations surrounding China's environmental investigative journalism.

The construction of agendas, such as pollution and resource exhaustion, generates a discourse of risk that defines what cannot be done for the sake of the environment, the knowledge of which contributes to consolidating the rule of the state by making institutions and individuals aware of the pressure of public opinion in action and therefore leads them to curtail their own activities. The fact that new markets are emerging, spurred by the understanding of environmental destruction, vividly epitomises not only the greedy nature of capital but also the logic of seeing technologies as the hope for survival which underlies consumer culture in our present society.

Investigative reporting on other issues such as dam and artificial islands/land construction, however, takes a critical stance towards certain policies of the state and represents diverse values prevailing in Chinese society. These values disapprove of the state's position in promoting economic activities and prefer a harmonious relationship between humanity and the environment. Reports of this kind mirror people's fear of environmental catastrophes caused by anthropogenic projects, challenging the dominant Promethean discourse lying beneath the practice of modernisation. The critical and advocacy function of investigative journalism, which contrasts with its ideological state apparatus role, is well embodied in the construction of this kind of agenda.

In terms of issues such as widespread cancers caused by environmental damage, investigative reports separate the groups of cancer victims who most of the time are based in rural villages from the rest of the population by labelling their living places "cancer villages". Though this helps to publicise their suffering and needs through powerful symbolism that arouses the general public's awareness of health hazards resulting from environmental factors and revealing environmental inequalities between rural and urban residents, such separation nevertheless may lead to negative consequences. For example, villagers who come from "cancer villages" may be socially isolated, and urban residents – especially from the middle and upper classes – may be convinced that such bad things will not happen to them as they live far away from such "cancer villages". This, on the one hand, indirectly helps strengthen the rule of the state by setting the minds of the middle- and upper-class urban residents at peace. On the other, this may impair the legitimacy of the regime since the majority of China's population live in rural areas.

The 20-year development of environmental investigative journalism suggests two important points concerning the interaction between investigative journalism and social contexts. First, it is as a result of intricate social dynamics that space has been generated for investigative journalism to report on environmental problems and issues. The successful publication of these reports captures well the complexity of social dynamics in China. The state is not monolithic. The decentralised government relationship leads the central government to be willing to constrain the activities of local governments. Although capitalism is at the centre of China's economic modernisation, the state has been aware of the threats capitalism may pose to the environment and the destructive consequences of unsustainable development. With commercialisation and globalisation accelerating, it is oversimple to see the state as monolithic and having absolute authoritarian control over Chinese media and society. It is also oversimple to see that the state and capital completely conspire against the environment. At this point, it seems the state well understands capitalism's exploitation of nature and would like to prohibit some of the activities of capitalism as well, though at the same time having a strong desire for economic development. Different government departments and officials have their own respective interests and take different stances towards environmental problems. Their diverse responses to and interpretations of environmental problems create room for investigative reports. Intense environmental protests and campaigns pressure governments and at the same time reflect the popular desire for a clean and safe environment. In addition, the frequent occurrence of natural disasters such as landslides, droughts, flooding and smog over recent years has triggered a nation-wide reflection on environmental risks.

Second, investigative journalism has captured opportunities provided by social dynamics, winning a relatively big space for reporting on environmental problems. As a consequence, investigative reports have offered a discourse space for diverse voices in society regarding environmental problems. Diverse or critical voices can be heard. The importance of environmental investigative journalism lies in strategically making use of the space created by social dynamics, constructing agendas of environmental risks at appropriate times and acting as a public arena in which not only active dissenters, but also those who are mainly silent and disadvantaged can make their voices heard. This is crucial in an authoritarian country that usually only allows the official dominant voices to be raised. Like other forms of communication, environmental investigative reports act as a platform that symbolically conveys the ideological meanings of certain issues to readers and as such has formed a new ideology of the relationship between modernization and the environment that is different from the dominant mainstream ideology.

3
The Discourse of Risk: Environmental Problems and Environmentalism in Chinese Press Investigative Reports

The previous chapter discussed the development of environmental investigative journalism over the 20-year period from the 1990s to the present and analysed the construction of nine agendas of environmental problems in the process and its implications for understanding the role of investigative journalism in mediating the rift between modernisation and the environment. Following the previous chapter, this chapter[1] takes the discussion further and offers a detailed account of the discourse of environmental problems as constructed in Chinese newspaper investigative reports. Discourse analysis[2] and framing analysis[3] of investigative reports on environmental issues in selected Chinese newspapers from 2008–2011 will underpin this account.

Discourse is something that defines what is meaningful and something that can transform power relations (Foucault 1972). Discourse is seen as crucial for exercising power and shaping power relations (Foucault 1972; Fairclough 1989, 1992, 1995). Discourse helps shape social structure, "social relationships between people", as well as "systems of knowledge and belief" (Fairclough 1989, 1992, 1995). In the case of environmental problems, the discourse about the environment has the potential to steer human societies' perception of their environment which influences their decisions about which way to go and what should be done. Media discourse of environmental problems provides insights into what has happened to the environment, shaping and fostering public knowledge and raising public awareness and anxiety about environmental risks. These insights impact on the environmental perception of social actors and their resulting actions. On the other hand, social actors, ranging from environmental victims to governments, struggle to influence the discourse about the environment for their own benefit and to get their stories and voices into news reports. Environmental discourse helps shape and is also intertwined with power relations within debates on environmental issues and policies. These debates and policies are often related to the development paths of societies. The importance of discourse thus drives social actors to exert control over the formation and reproduction of the discourse of the environment. One crucial site where the ongoing contest among social actors takes place is in the coverage of the news media, in which the discourse of the environment is constructed.

Media construction of environmental problems, of course, does not mean environmental problems do not exist but are invented by the media. Instead, it merely refers to the particular ways in which media coverage interprets environmental problems, especially the definitions, causes and consequences of environmental problems. Media interpretation is influenced by and influences social dynamics and tensions in the external social context. Cancer villages, for instance, existed in China long before the mainstream media started reporting on them in the late 1990s. The term "cancer villages" emerged in China's public arena only after the publication of the first investigative report exposing this issue in *Beijing News* in 1998. The emergence of this risk to health resulted from modernisation, but only became part of the public agenda when the problem became severe.

This chapter deconstructs the way in which environmental problems are presented and interpreted by investigative reports. This contributes to our understanding of the formation of environmental discourses in the texts of investigative reports. This chapter also discusses the meaning of the discourses and evaluates the role they play in China's modernisation. Framing analysis and discourse analysis of the 258 investigative reports reveals that there coexist two discourses of environmental problems in these reports: a tragic discourse of extinctionism and a radical discourse of eco-equalism. Marxist environmentalism emerges in the two discourses. The discourses express worries over the environment of China and over the health and safety of local people and the extinction of species. The anxieties come out of the understanding of the inevitable, negative impact of economic activities and modernisation on the environment. The road to modernisation appears to lead to an unavoidable disastrous fate for China. These discourses of environmental problems are characterised by a dominant narrative that attributes environmental deterioration to human activities, especially economic activities. This narrative reflects a critique of the destruction of the environment caused by state capitalism. Human beings are portrayed as competing for resources with one another and with other species, while governments and capital are depicted as being hand in hand in the destruction of nature. These reports blame human beings for their shameless exploitation of natural resources. In particular, socio-economically advantaged people are exploiting natural resources but leaving disadvantaged people to bear the consequences of such exploitation. There is a close link between social and environmental inequalities. Environmental deterioration is thus inevitable, as a result of the greed of humans for capital accumulation. The greedy nature of capitalism and human beings is an enemy of nature. In a word, the discourses of these investigative reports are warning that human beings' activities will eventually kill us all and other species on earth if nothing is done about environmental problems and hazards.

Through reflecting on what has happened and is happening to the environment, the two discourses oppose the top-down and long prevailing Promethean discourse about the relationship between the environment and modernisation. They potentially attempt to cool down the enthusiastic lust for economic development initiated by the central government in Beijing and later evident across the country. The two discourses form and exercise a soft, pervasive, capillary and evolutionary power that aims to prompt individuals to sound the alarm about environmental problems when they become aware of the worsening environment and the possible consequences they may bear. This facilitates the development of "reflexive modernisation" in China through providing an alternative space for the general public to reflect on what is happening to the environment.

The rest of this chapter will introduce two types of discourses that have been identified in the texts of newspaper investigative reports, followed by a discussion of the meaning of the two discourses while considering its relationship to recent social dynamics regarding environmental problems and modernisation.

3.1 Sick environment and collective fear: a tragic discourse of extinctionism

These newspaper investigative reports construct a tragic discourse of extinctionism, characterised by a collective fear of environmental crisis. The environment is depicted as sick and suffering from the effects of human activities. Seven types of environmental problems are defined and framed in the ten newspapers' investigative reports over four years

Table 3.1 Background information of the ten newspapers selected for framing and discourse analysis *

Newspapers	Types	Locations	Years of Publication	Circulation (daily)	Editorial policies (or slogans)
People's Daily	Party Organ	Beijing	1948	2.3 millions	To communicate the Party's policies and important information to the people
Beijing News	Commercial	Beijing	2003	776,000	To independently and objectively report current affairs and politics; to shoulder social responsibility
Beijing Youth	Commercial**	Beijing	1949	600,000	To report the world from a youth's perspective; to have opinions, depth, taste and social responsibility
First Economic Daily	Commercial	Shanghai	2003	800,000	To be responsible for the era; to provide authoritative, professional, rational, and responsible (financial) news
Oriental Morning	Commercial	Shanghai	2003	400,000	To objectively and independently communicate economic information in the Changjiang Delta area
Southern Metro Daily	Commercial	Guangdong	1997	1.83 millions	To record contemporary social development, nurture the emergence of modern society, and enlighten civil consciousness

Southern Weekend	Commercial (weekly)	Guangdong	1984	1.7 millions (weekly)	To provide an understanding of China and offer reports that reflect justice, conscience, love, and rationality
Yunnan Info Daily	Commercial	Yunnan	1985***	380,000	To report political and economic issues; to have international influences
Xiaoxiang Morning	Commercial	Hunan	2001	602,000	To give voice to the people, help maintain social justice and report the truth
Dahe Daily	Commercial	Henan	1995	700,000	To care for and serve the people and be close to people's life

*Data collected from the websites of these newspapers and that of Phoenix TV in March 2013
** *BY* is owned by Beijing Municipal Committee of China Communist Youth League, but is operated commercially and has a tradition of critical investigative reporting.****Southern Daily* Press Group took over part of the ownership in 2007. Since then, the newspaper started practicing investigative journalism.
Source: Tong 2014.

Table 3.2 Types of environmental problems represented in ten newspapers' investigative reports during 2008 to 2011

Types of environmental problems	Frequency	Percent
Pollution and pollution-caused health problems	167	64.7%
Human-induced geological problems and resource shortages	63	24.4%
Ecological crisis	10	3.9%
Multiple environmental problems	10	3.9%
Climate change and global warming	3	1.2%
Extinction of species	3	1.2%
Transgenic food	2	0.8%
Total	258	100%

from 2008–2011 (illustrated in Table 3.2). The nine agendas discussed in the previous chapter have been integrated into the seven types of environmental problems. These environmental problems, ranging from pollution to resource shortages, are all caused by the activities of human beings in the modernisation process rather than being naturally occurring. Even if the reports are about natural disasters, such as landslides or earthquakes, the environmental problems concerned are still framed as an anthropogenic issue.

There were minor changes in the discourse of environmental problems over time (see Figure 3.1 and Table 3.3). Despite the changes, over the years "pollution" remains at the top of the list of all environmental problems, while "human-induced geological problems and resource shortages" increasingly became pressing topics that are seen as generating risks for China (see Figure 3.1). In 2008 the coverage of environmental problems was very simple, comprising only two categories of environmental problems: "pollution and health problems caused by pollution" as well as "multiple environmental problems". From 2009 onwards, new types of environmental problems appeared, such as "human-induced geological problems and resource shortages", and associated "ecological crises". Overall, "pollution and human health problems caused by pollution" (64.7%) and "human-induced geological problems and resource shortages" (24.4%) are the two main environmental problems among them (see Table 3.2).

"Pollution" (152, 58.9%) has been the number one environmental problem haunting China. From pollution incidents to general pollution problems, these investigative reports have depicted a terrifying picture of China's environment. China's rivers have been poisoned because of the discharges by chemical or other industrial factories. Investigative reports record a series of water pollution incidents across China, including water pollution in Yancheng City, Jiangsu Province, arsenic poisoning of the

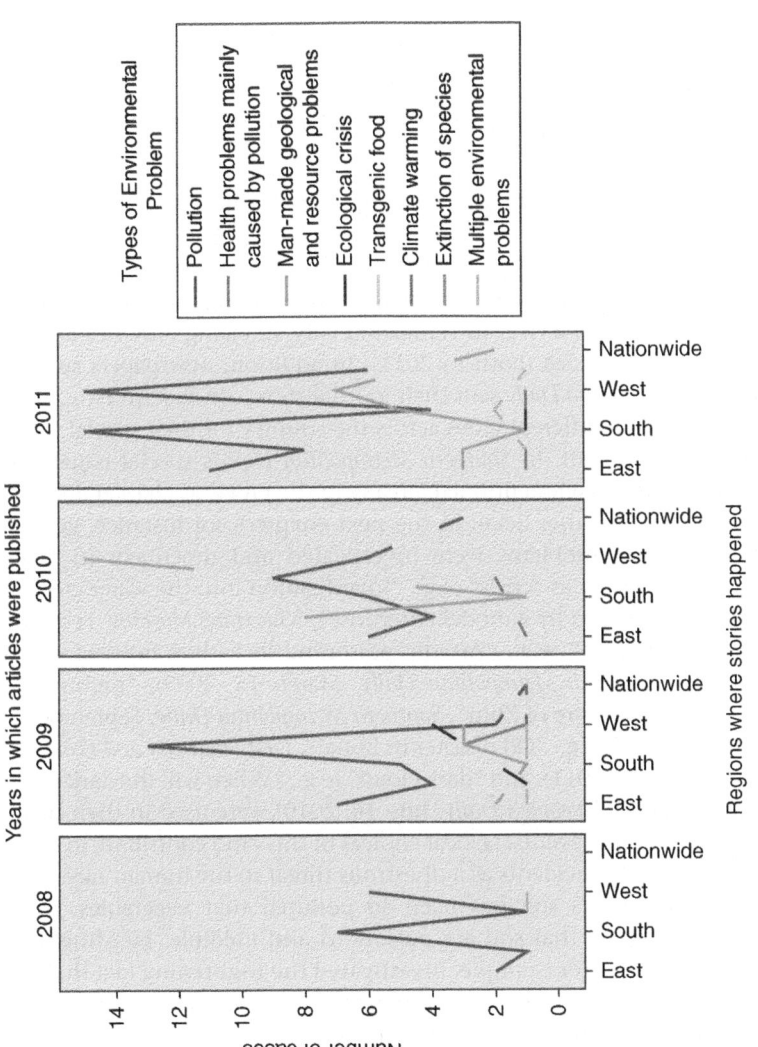

Figure 3.1 Geographical distribution of coverage classified by the types of environmental problems over time

Table 3.3 Average appearance of each frame per year

	2008	2009	2010	2011
The consequence frame	2.52	2.81	2.82	2.73
The conflict frame	2.65	2.76	2.33	2.36
The responsibility frame	2.15	2.47	2.41	2.41
The human-interest frame	2.19	2.38	2.27	2.21

Scale from 1.00 (frame absent) to 3.00 (frame present) (Dirikx and Gelders 2010)
Source: Tong 2014.

river in Minquan County and Shangqiu City, Henan Province and the muddy drinking water issue in Nanzhang City, Hubei Province (all three in 2009), the Zijin pollution event in Shanghang County, Fujian Province, water pollution in Chifeng City, Inner Mongolia (both in 2010), phenol pollution of a river in Hangzhou City, Zhejiang Province and oil leaks in the Bohai Sea (both in 2011). In addition, newspapers such as *Southern Metropolitan Daily* sent their journalists to conduct investigations into pollution in different cities across the country for their special issues on the environment. In *Southern Metropolitan Daily*'s special issue (this special issue has been mentioned already in the previous chapter and will be discussed in further detail in the next chapter), for instance, general water pollution problems were investigated and described. In these reports, words such as "crisis" (e.g., "Investigation into the water crisis in Yancheng: poisoned by a model enterprise", *Xiaoxiang Morning*, February 24, 2009), "poison" (e.g., "Arsenic poisoning in a river flowing across provinces", *Southern Metropolitan Daily*, March 15, 2009), "nightmare" (e.g., "The Nightmare of Zijin", *Southern Metropolitan Daily*, September 1, 2010), "disaster" (e.g., "Oil disaster in Bohai", *First Financial and Economic News*, August 10, 2011), and "dark cloud" (e.g., "When will the dark cloud of Zijin disperse?" *People's Daily*, July 16, 2010) were used in their headlines to refer to the events. Lexical choices of this kind contribute to interpreting pollution incidents as a disastrous threat to the human race.

Much of China's soil has been so polluted that vegetables, fruits and rice grown in that soil are poisonous and inedible. Lv Minghe at *Southern Weekend*, for example, investigated the frightening fact that the rice supplied by a national rice producer in Hunan Province had been polluted by heavy metals as a result of being grown in soil contaminated by them. However, the polluted rice had been sold to consumers ("Heavy rice", January 6, 2011). His article raised concerns over food security in China, criticised industrial activities for polluting the soil and blamed some merchants' avarice and lack of conscience. While the soil becomes infertile, rivers and lakes face a similar fate. As such, diverse species of fish either die out in rivers because of pollution or vanish due to over-fishing. Peasants have no harvest to celebrate and no agrarian products to sell. Fishermen have no fish to catch in rivers and lakes. The investigative report by Pan Yibing and Chen Jun, for instance, looked into the reality that man-made environmental problems, such as pollution and sand mining, had destroyed the quality of water and riverbanks, which led to the rapid decline of fish numbers and species in rivers and thus fishermen were facing a jobless future ("Sticky water, less fish, 2,000 fishermen have to take days off," *Southern Metropolitan Daily*, March 23, 2011). These pollution problems appear not only in heavily commercialised coastal cities but also in developing inland regions. These problems have spread across China, expanding from eastern seas to western deserts and extending from northern forests to southern rivers.

Geological problems and resource shortages appear as a consequence of two types of human activities: anthropogenic projects, such as dam construction and land reclamation, and the over-exploitation of natural resources, such as mining and logging. Both types of human activity have a long history in China. A fever of dam and other hydroelectricity station construction has swept China since the establishment of the PRC, as discussed in Chapter 2. Numerous dams and stations of this kind were established during the Mao era from the 1950s to the 1970s. The wave of dam construction in the 21st century is merely a continuation of Mao's passion and ambition. Not only the hydroelectricity stations built in the past, but also those constructed in the new century, have been reflexively reviewed in some investigative reports. Meng Dengke's investigative report in *Southern Weekend* ("The calamitous consequences of a thousand hydroelectricity stations on Jiulong River", April 28, 2011), for example, argued that the construction of about a thousand hydroelectricity stations in the Jiulong River region had led to a severe geological and ecological crisis.

The same is true of the over-exploitation of natural resources. An investigative report written by Peng Liguo published in *Southern Weekend* in 2011 ("Northern Barn will be gone with the loss of black soil", January 6, 2011), for example, revealed a serious problem the northern part of China, historically seen as the great grain-growing area, is facing: the loss of black soil due to human-induced activities such as over-logging. This report warned that most of the black soil that is essential for growing grain will be gone in 40 to 50 years if we do nothing to stop it from happening.

Capitalism and the political authorities have become two dominant forces that degrade the environment. Capitalism's greedy nature and Maoism's motto of "men conquer nature" have driven coastal governments to exploit sea resources and inland governments to profit from underground resources. Sea regions suffer from reclamation activities launched by coastal governments (see "No quiet land in ten thousand coastlines" by Dengke Meng and Qing Zhang, *Southern Weekend*, March 31, 2011, for example). Thanks to unsustainable exploitation of these natural resources, regions that used to be proud of their underground resources such as copper and coal are now facing a crisis of resource depletion and exhaustion which has even resulted in a hollowing process underground. In 2009 a series of investigative reports published by *Beijing News*, such as "Daye, the exhausted copper capital" (March 18, 2009) and "Buxin after the depletion of hundreds of miles of coal" (March 20, 2009), accurately documented this crisis.

Associated with these are human health problems, ecological crises and extinction of species. Investigative reports on "human health problems" (15, 5.8%) mainly reveal two types of problems: "cancer" (6, 2.3%) and "high levels of lead poisoning" (4, 1.6%). In the accounts of investigative reports, cancers are depicted as a nightmare critical disease caused by environmental deterioration, in particular pollution, and also a collective disease that everyone living in particular villages may suffer from. In other words, cancers are interpreted and framed as an environment-induced issue instead of a health problem caused by other factors such as genes and lifestyles. In some cases people in the villages where most of the population was affected were afraid of pollution and pleaded with governments to remove the pollution sources – local factories or mining sites. Pollution also explains the suffering of children from lead poisoning in different regions such as Shanxi Province, Shanghai, and Hunan Province.

Seas, famous rivers and lakes, such as the Great River, the Panyang Lake, the Dongting Lake, numerous other rivers and lakes, and even underground water face problems of pollution and low water levels. Some are even disappearing altogether. Accompanying this ecological crisis is the extinction of species, such as the finless porpoise and the baiji (a freshwater dolphin found only in the Yangtze). For four years *Southern Weekend* and *Southern Metropolitan Daily* did investigations and covered the sad stories of the disappearance of finless porpoises and baiji (such as "The tragedy of baiji: end or beginning?" and "The survival crisis of cave fish" in *Southern Weekend*, respectively on January 21, 2010 and August 6, 2009; "Will finless porpoises in the Yangtze inevitably repeat the tragedy of the baiji?" in *Southern Metropolitan Daily*, on January 26, 2011). All in all, China is portrayed as a land not suitable for living in now either for humans or for animals. This is a picture of catastrophe from which no one in that environment can escape.

Meanwhile the discourse of extinctionism is characterised by a collective fear. This is first of all reflected in the lexical choices of these investigative reports. A frequency analysis of words used in these investigative reports shows that "pollution" was the word that was most frequently used. It appeared 2,866 times in these 285 investigative reports. The top ten most frequently used nouns include "pollution" (2,866 times), "ill" (2,255 times), "environment" (1,932 times), "government" (1,898 times), "chemistry and industry" (1,685 times), "enterprises" (1,685 times), "environmental protection" (1,664 times), "China" (1,631 times), "problems" (1,476 times) and "villagers" (1,368 times). These top ten most frequently used nouns suggest an emphasis on and a relationship among pollution, illness, government, chemical/industrial enterprises, villagers and the homeland.

Negative words, associated with death and crisis, were heavily used throughout the reports and across media. In the 258 investigative reports, for example, the word "death" was used 1,217 times, the word "poison/poisoned" 1,253 times, "polluted/pollution" 2,866 times and "cancer" 717 times. Expressions associated with crisis and fear are also among the most frequently used words and phrases, such as "crisis" 174 times, "problem" 1,476 times, "risk" 260 times, "worry" 80 times, "disaster" 49 times, "fear" 151 times, "panic" 87 times, "chaos" 32 times, "dreadful" 20 times, "victim" 203 times, "grievance" 98 times, "nightmare" 10 times. The use of these negative words indeed weaves an image of crisis and risk in China.

A picture of collective fear is depicted also through describing the consequences of environmental problems and the despoiled environment itself. A close look into the content describing the consequences of environmental problems and the situation people are living in reveals that investigative journalists have often made lexical choices, such as "death", "ill/illness", "crisis", "fear", "upset", "risk", "disaster", "threat", "conflict", "finished", "extinction/disappear", "grievance", "shadow", "chaos" and "distrust", to describe the environment and people's feelings about and responses to the environmental problems they are facing. Words such as "poisoned" (appearing 1,253 times), "barren" (439 times), "black" (557 times) and "stinking" (295) were frequently used to describe the environment investigative journalists saw and experienced in their reports. In addition, investigative reports are inclined to cite official claims and scientific knowledge to increase the authority of their reports and quote journalists' own witnessed scenes to describe the terrible situation. For example, an investigative report in *Beijing News* (How can river water poisoned by Arsenic in Minquan, Henan flow through four cities without being noticed by four local environmental monitoring departments?" March 24, 2009) describes the possible consequences of the accidental arsenic poisoning of a river by citing a government environmental protection officer's comments in the following words "if the polluted water flows into the urban water system in the cities of Haozhou, Buyang, Benbu that are located in the downstream of the river and caused arsenic poisoning, we will face disastrous consequences" and followed by a definition of "arsenic poisoning" and an explanation of its effects: " 'Arsenic Poison' is 'White Arsenic Poisoning' (*pishuangzhongdu*).[4] If arsenic invades human bodies, it will cause skin keratinocytes, canceration, systemic chronic poisoning, and eventually death." The journalist also described the scene in these words, "the Dasha River was a dark brown colour and had a constantly stinking smell, showing a decaying scene", which depicted the terrifying scene of the polluted river.

These discursive strategies can be summarised in an investigative report entitled "People's psychological fear of chemical factories behind a rumour" (February 21, 2011). In this report, Yang Xiaohong from *Southern Metropolitan Daily* depicted an image of collective panic and fear prevailing among local residents living near the chemical industry zone of Chenjiagang Town, Jiangsu Province. Such a fear led some tens of thousands of local residents to believe a chemical factory was going to explode and therefore they fled for their lives, leaving their homes on a snowy night. At least four people died in their escape. An old man's words were quoted in the report: "We are living in a powder keg. If you were me, wouldn't you be afraid?" The context discussed in the report shows that the collective fear of local residents is justified by background information, such as a previous chemical factory explosion, a chlorine leak from a chemical factory, several chemical poisoning accidents in the past, pollution, the decline of local fishery and agriculture caused by chemical pollution, the extinction of stellar whales and the threat to red-crowned cranes who yearly migrate here to spend the winter, as well as local residents' illnesses and the number of local cancer victims which together were referred to as "dangerous memories of the chemical zone". The journalist uses discursive strategies to imply her meaning. When discussing the number of people who died in the explosion, she wrote, "according to the then official statistics, eight people died and dozens were injured in this explosion. But the names on the dead list were all names of local residents". Since it is common knowledge in China that most workers who work in chemical factories are migrant workers who come from outside rather than local residents, this sentence implies that the number killed in the explosion as reported by the local authority may not be accurate and may be lower than the actual number. Words and phrases, such as "strong unpleasant smell throughout the year" and "increasingly dirty and muddy river", are used to describe the local environment. The consequences of local environmental problems caused by chemical factories in the zone included the following examples: "last year seven or eight hundred little wild duck died at once", "two or three Mu of Chinese cabbage died overnight", "189 domestic rabbits and ducks gradually died", "five or six people died of cancer last year within a 500-meter radius of Caogang Village", "76 villagers died of lung cancer, stomach cancer or other types of cancer from 2000–2005 in Dongjin Village", "most fish and shrimp disappeared along the coastline", "all shellfish died", "the skins of fishes caught by fishermen had already rotted. If you cut open some fish, you would find only water inside the fish's stomach (no food would be there)", and "in 2008, the environmental protection department in Xiangshui County tested the water in the Guanghaikou estuary which showed the water was 'inferior v type' surface water that was not suitable for fish to live in". From this example one can see that a collective fear for the sick environment and environmental hazards has been constructed in the report through the adoption of discursive strategies such as the choice of words, descriptions of background information, quotes from victims and villagers, and citation of official statistics.

3.2   (State) Capitalism, class and destruction of nature: a radical discourse of eco-equalism

While China's political leadership keeps promoting the belief in "humanity defeating nature" among its people, Chinese investigative reports tell a completely different story for their readers. The stories have drawn a clear binary line between economic development/human activities and the health of the environment. Investigative reports follow a repeated narrative in which not only are economic activities encouraged by the priority given by Chinese governments to economic growth and modernisation, but human activities in general are to blame for environmental problems appearing in China. The human race and other species in the world will be punished in turn by nature, which has been

Table 3.4   Average appearance of each frame

The consequence frame	2.76
The conflict frame	2.33
The responsibility frame	2.40
The human interest frame	2.27

Scale from 1.00 (frame absent) to 3.00 (frame present)
(Dirikx and Gelders 2010)
Source: Tong 2014.

damaged by human beings. This is the pessimistic and tragic discourse of extinctionism which has been discussed in the previous section. This discourse shares the concerns of the Dryzek survivalism discourse that sees resources as limited and finite and the relationship between humanity and nature as conflicting (Dryzek 2005). But this discourse has a different view about the role of elites and governments, who are regarded as untrustworthy due to their selfish nature, that is, the involvement of elites and governments will only lead to an ever larger gap between socio-politically advantaged and disadvantaged people and social groups. This is the radical discourse of eco-equalism which will be discussed in this section.

Associated with the discourse of extinctionism is a radical discourse of eco-equalism that accuses capitalism and authoritarian governments of destroying nature and increasing inequality in society. This is, first of all, reflected in an overall emphasis in newspaper investigative reports on the causes and consequences of environmental problems. This emphasis can be seen in the average appearance of four frames (see Table 3.4) analysed in this study. All four frames appear close to the end of the "frame present" end of the continuum, that is to say, a continuum from "frame present" at one end to "frame absent" at the other. The two frames – consequence and responsibility – have a slightly higher appearance than the other two frames. The appearance of these frames has not changed much over the four-year period. From Table 3.3 one can see that both the responsibility and consequence frames from 2009–2011 are more prominent than in 2008.

This emphasis on causes and consequences is also supported by the frames of conflict and human interest. In these investigative reports, three types of conflicts – conflict between environmental problems and economic reform (41.4%), conflict between the victim(s) of environmental problems and commercial organisations (19.6%) and conflict between officials/government and ordinary people (16.5%) – are portrayed as the primary conflicts associated with environmental problems. The majority (76%) of investigative reports also describe how individuals have been affected by environmental problems. Adjectives have been used in the descriptions of the environment in most of the investigative reports (85.7%), with a potential to arouse compassion among readers about environmental deterioration and environmental victims' suffering. In the reports of this kind, a clear causal relationship is established between pollution and industrial activities. Investigative journalists tend to attribute pollution problems to industrial activities. For example, in Li Jun's investigative report titled: "Granary in Guanzhong: cities stole and cut the water for agriculture" (*Southern Metropolitan Daily*, February 16, 2009), she investigated the causes of severe drought in Northern China, especially in remote areas, and concluded in her report that the over-rapid needs of urbanisation bear the main responsibility for the severe drought which explained how this had happened.

A close qualitative examination of these reports reveals a more detailed picture of how they have portrayed who should be held accountable and who are the environmental victims. Economic development is portrayed as the number one enemy of nature. Two actors – governments and enterprises – are held responsible for environmental problems. The reports mainly blamed the activities of commercial organisations, such as chemical factories, and individuals such as coal mine owners (37.9%), government policies (28.9%) and the needs of industrialisation and modernisation (21.2%) for the deterioration in the environment. Surprisingly, climate change and global warming (0.6%) is not portrayed as a major cause of environmental problems. For most of the time government policy and enterprises' activities are equally blamed. Meanwhile, agriculture and fisheries suffer from industrialisation and the national drive for economic development. Agricultural pollution, for example, is not accredited to the application of chemical fertilisers in agriculture. Instead, agricultural products are contaminated by the waste discharged by industrial factories. Workers, farmers, peasants, fishermen and herdsmen (32.8%), and local residents (22.8%) – as well as the ecosystem and nature itself: rivers, oceans, the soil and so on (32.1%) – are seen as the main victims of environmental problems.

The reports depict an image of indifferent and incompetent local governments. It is local governments that strongly support the construction of chemical factories through policies that reward enterprises for doing so. Among the 258 investigative reports, most imply that governments should be responsible for environmental problems and that there were conflicts between governments and the environment and

Table 3.5   Conflict types mentioned in the investigative reports

Conflict types	Number of appearances	Percent
Conflict between environment/victims and government (policies and officials)	277	66
Conflict between environment/victims and commercial organisations/individuals	89	21.2
Conflict between commercial organisations/ individuals and government	14	3.3
No conflict	14	3.3
Conflict between industry and agriculture/fishery	12	2.8
Conflict between different points of view/sets of data	5	1.2
Conflict between different regions/levels of government regarding environmental problems	5	1.2
Conflict between human activity and environment	2	0.5
Conflict between rich and poor	2	0.5

environmental victims. Only 7.4% do not mention that governments should take the responsibility for environmental problems. Only 6.2% of reports didn't mention conflicts between two or more parties. Of those reports that do mention conflicts, many of them refer to more than one type of conflict. Conflicts are mentioned 420 times in these reports. From Table 3.5, one can see that conflicts between environment/victims and government are most frequently cited – 277 times (66% of reports). The second-biggest conflict lies between environment/victims and commercial organisation/individuals with 89 appearances (21.2% of reports). It is government and its economic associates that are in conflict with the environment and environmental victims and should be held responsible for environmental problems. This matches the explicit analysis of who should be responsible for the environmental problems offered in these reports. This analysis accuses "Government policies, priority for economic growth and politicians" and "Commercial organisations/individuals" of being two culprits who should take the responsibility for the deteriorating environment (see Table 3.6).

For example, in "Investigation into the water crisis in Yancheng: poisoned by a model enterprise" (in *Xiaoxiang Morning*, February 24, 2009), a chemical factory was blamed for having polluted the water system in Yancheng City that led to a water crisis. The author further sarcastically revealed that local government had awarded this chemical factory "model enterprise" status for its protection of water sources seven months previously, just before the occurrence of the environmental accident. The context of this accident explained in the report described a

Table 3.6   Who should be held accountable for environmental problems?

Who should be held accountable for environmental problems?	Number of appearances	Percent
Government policies, priority for economic growth and politicians	250	50.1
Commercial organisations/individuals	189	37.9
Local residents	21	4.2
Whole population/human beings	18	3.6
Natural environment	7	1.4
Not applicable	6	1.2
Scientists	5	1.0
Global warming	3	0.6

picture of local government supporting and protecting this chemical factory, and even encouraging the operation of factories of this kind due to its concern to increase local revenue, but, meanwhile, turning a deaf ear to local residents' complaints, appeals and losses. The environmental problems started with people finding that crops and fish were dying, resulting in financial losses from which farmers suffered, and this became more severe when later ordinary residents in the whole city fell victim.

This narrative and its structure, which has become a reporting mode, or even a paradigm, can be easily identified in many other newspaper investigative reports on environmental accidents and problems examined in this study. Such a paradigm directly and indirectly targets governments – either local governments or the central government – as to blame for environmental problems. This point is prominent in the close association between the construction of man-made projects and pressing environmental problems. Anthropogenic projects are also blamed for the appearance of environmental problems and the deterioration of ecology.

For example, investigative reports on the environmental consequences of the Three Gorges Dam suddenly appeared in 2011. We have six investigative reports on this topic published in 2011 in the 2008–2011 sample. These reports clearly pointed out that the deterioration in ecology in the Three Gorges area was caused by the construction of the dam and hydroelectricity stations. Though both the government and hydroelectricity developers should be blamed for this, overall it was the central government that decided to launch the projects and thus should take the responsibility. Some of these reports used natural disasters as a starting point to tell the stories. For example, both "Abnormal environmental phenomenon in the region of Three Gorges" (in *First Financial and Economic News* on May 31, 2011) and "'Roasting' Three Gorges" (in

Southern Metropolitan Daily on June 2, 2011) started their investigation into and told the stories about natural disasters, such as drought and low water levels in rivers, suggesting these were the environmental consequences and costs of Three Gorges and that governments should take the responsibility.

Therefore the discourse of environmental problems in investigative reports highlights that conflicts exist between humans and animals as well as between advantaged and disadvantaged people and social groups when explaining environmental problems and their causes and consequences. The discourse of environmental problems attributes this conflict relationship to the sin of capitalism. If using the idea of "anti-ecological" (Layfield 2008) here, capitalism is described as not only constantly destroying but also resisting pressures to stop the destruction of the environment. Capitalism's ignorance of environmental principles is reflected in its over-exploitation of natural resources which leads to the shortage of natural resources across the country that has been discussed above. The violation of environmental principles by capitalism is worsened rather than mitigated by the collaboration between capitalism and the state in the form of state capitalism. In other words, the priority for economic growth lures Chinese governments, especially local governments, to tolerate or even cover up capitalism's violation of ecological principles.

An important part of the narrative paradigm is the link between environmental inequality and social inequality constructed in these investigative reports. This link is an extension of blaming economic development for causing environmental problems. The detailed interpretations of the causes and consequences of environmental problems link them to social injustice and inequality. Such an interpretation constructs a dichotomy of politically and socially advantaged versus disadvantaged people or institutions or regions, producing a crisis discourse of the current capitalist mode of production and relevant policies. As discussed above, for example, environmental victims include agriculture, farmers, fisheries, fishermen/women, workers, nature, animals and vegetation, whereas capital, entrepreneurs and governments are environmental problem generators. Besides, this is not only about class inequality but about geographical inequality. There is an interesting geographical distribution of environmental problems in the coverage which provides more evidence for the link to social injustice and inequality – in this case, regional inequality. In fact, the chi square value (=.000, significant at the 0.05 level) suggests a statistically significant relationship between the types of environmental problems and the regions which suffer from these problems. The Eastern and Southern regions that were the first to launch the economic reform programme and have benefitted most from economic development are not the areas that have suffered most from environmental problems. Only around 31% of investigative reports are about environmental problems happening in these regions. Instead, the Western, Midland and Northern regions that have lagged behind in terms of economic development are the regions that have seen the worst environmental problems. About 62% of the coverage portrayed environmental problems in these regions. Figure 3.1 shows that investigative reports on environmental problems in the Western regions dramatically increased in 2010 and 2011. Similarly, the number of reports of environmental problems in the Midland regions suddenly became greater than before in 2009, though overtaken by attention focused on the Western regions from 2010.

This radical discourse of eco-equalism therefore sees environmental problems as problems with the nature of human beings and with society rather than problems with nature. This discourse shows distrust in, and disapproval of, elites and governments that are likely to exploit resources for their own benefit. A discourse of this kind is a reflection of class struggle between people and struggle among humanity and nature. This is a class-struggle perspective on the environment that uses equality to assess environmental problems. It is about the socio-economically advantaged social groups' exploitation of disadvantaged social groups. This meanwhile reflects investigative reporters' desire for equality in society and in the living environment for humans and for other species.

These points were confirmed by interviews with journalists. In the interviews a majority of journalist participants regarded economic reform as the main cause of environmental problems and pollution as the most prominent environmental problem. For example, a journalist explained the media frames in the reports in the following words: "the causation (between pollution and economic development) emerges quite naturally because pollution must have some relationship to economic development" (Interview 2011). Another explanation of journalists' perception about environmental problems is the notion of the exploitation of politically and materially disadvantaged people by advantaged groups. The following comments made by a journalist are very representative of this argument: "basically this is about the shameless exploitation by a few economic supermen/women over the living space and resources of the majority of people who have no power and chance to choose where to live. This is a problem about humanity rather than a problem about nature" (Interview 2011).

Viewpoints of this kind are prominently exemplified in many investigative reports, including those cited in the discussions in this chapter, such as "Heavy rice", by Lv Minghe, (*Southern Weekend*, January 6, 2011), "Will finless porpoises in the Yangtze inevitably repeat the tragedy of the baiji?" by Yang Chuanmin (*Southern Metropolitan Daily*, January 26, 2011), "Landslide in Linxiang: natural disaster or human disaster?" by Lv Minghe (*Southern Weekend*, June 20, 2010) and "Minqing in Gansu: the oasis is dying in sand storm", by Yang Xiaohong (*Southern Metropolitan Daily*, July 27, 2009).

3.3 The implications of the two discourses

The two discourses discussed above suggest the consequences of environmental problems are socially rather than naturally constructed. Notably, they embody prominently Marxist environmentalism to a great extent. The Marxist environmentalism in these newspaper discourses has three features: concern for environmental problems; the relationship between human and nature; and the consequences of environmental problems. First, capitalism is seen as a rival of the environment. Capitalist activities will inevitably lead to the destruction of nature. Economic activities and governments' decisions are held responsible for the sick and even catastrophic environment. The conflict between economic activities and nature always exists. Secondly, they depict a clear picture of class struggle between socially and economically advantaged and disadvantaged people, social groups and regions. The former exploit natural resources while the latter bear the consequences of such exploitation. The reports suggest those people or regions that occupy worse socio-economic positions suffer more from environmental problems. Third, the state and local governments have been found to be unable to deal properly with environmental problems, given their capitalist interests, poor governance, and technological and management incompetence. The reports suggest the mutual interest–based collaboration between local governments and powerful commercial organisations has created barriers for environmental protection and undermined the efforts of the central government in promoting ecological modernisation.

The three aspects of environmentalism in the newspaper discourses to a great degree fit the arguments of Marxist ecologists that nature falls into the object of capitalist projects and inevitably has conflicts with capitalism (Nayar 2010; Gare 1995). In his writing, Marx explains his views of nature and the relationship between human society and nature. He suggests nature provides raw materials for human labour and is exploited and jeopardised in the capitalist mode of appropriation and by the capitalist class relationship (Marx 1976). Marx sees the resource shortage problems as caused by the ineffectiveness of capitalism (Perelman 1996). Environmental crisis is seen as resulting from "the specifically capitalist form of organisation of economic life" (Benton 1996: 7). Discourses with these features reflect these views and also echo the arguments raised by Marxist scholars who see environmental problems as impacting on people disproportionately depending on their social status and are always linked to social inequality (Layfield 2008; Boucher 1996; Gare 1995). Social inequalities occur when the environment as the object of the labour process is used and exploited in the process of production that facilitates the exploitation of one class over other classes. In addition, the discourses also suggest that to a certain degree no one can escape the consequences of environmental degradation by portraying the disastrous and sick environment that exists everywhere. That is to say, no matter which class a person comes from, no matter which position a person occupies in society, environmental problems are so pervasive that s/he cannot escape from them. This is what Marx argues: we are all members of the "environmental proletariat".

While accepting Marxist environmentalism in terms of class struggle and the anti-environment nature of capitalism on the one hand, nevertheless the discourses disagree with Marx in terms of "men conquering nature" on the other. Marx had confidence in the ability of men to control and conquer nature through technology rather than being controlled and enslaved by it. As discussed in Chapter 1, the Chinese leadership has inherited this idea from the outset. This was exemplified in sequential social movements that were initiated from the top down with the aim of conquering nature. The Great Leap is a prominent example of this. The economic reforms since the 1980s continue this type of thinking, and that is integrated into both the letter and implementation of relevant economic policies. The Marxist and Maoist ideology of conquering nature inspires governments at all levels to make the most extreme use of natural resources to meet their perceived need for economic and political achievements. This is precisely one of the crucial things these reports are against.

Following Foucault's understanding of discourse, these two discourses of environmental problems contribute to the production of knowledge about the environment and modernisation. This kind of contribution is particularly important in a society where the dominant discourse about the environment is the Maoist belief that "men conquer nature". The two discourses however construct a discourse of risk that implies the state and capital are the two dominant factors that lead to environmental deterioration. In addition, environmental risks arising in the process of modernisation are leading ordinary Chinese residents towards a dismal future rather than a better life, although that is supposed to be the ultimate aim of modernisation.

Such anti-capitalist discourses achieve some justification from their particular social context and indeed are part of the public discourse of the environment and modernisation that is currently prevailing in society. The appearance of the two discourses has not been sudden. Instead it has a context and a dialectical interaction with dynamics in society. In recent years various agendas have entered public discourse about the environment. Environmental agendas, for instance, are included in the coverage of the national press[5] and have even been set by the increasing emergence of environmental protests. National newspapers have debated multiple topics regarding environmental problems over a twelve-year period from 2000–2012. Among others, rubbish-burning, cancer villages, the Three Gorges Dam, and big chemical projects (*dashihua xiangmu*) are four of the most popular environmental agendas in this period. Judging from newspaper space given to the topic of rubbish-burning, for instance, one can find there has been a long-term attention to this subject, with a sharp increase in 2009, remaining high from 2009–2011. Over the ten years 1,748 articles on this topic were covered by 166 newspapers. Similarly, the ten-year trend of newspaper attention to the topic of cancer villages maintained stable from 2000 but increased suddenly in 2011. The topic of the Three Gorges is a bit different. The most intense newspaper attention was seen in 2003. Then newspaper coverage declined rapidly. However, examining the tone of the 2,646 articles on the Three Gorges from 2000 to 2012, one can find only 190 articles that expressed a critical attitude towards the construction of the Three Gorges, and raising questions about the impacts and consequences of the project to the environment. The year that saw the biggest number of negative articles (57) was 2005 and 2011 saw the second-biggest number of negative reports (34 articles). The year when newspapers carried the biggest number of articles on big chemical projects was also 2005. This year saw the first negative reports about big chemical projects. Despite the fact that such projects have been launched across China at an increasingly rapid pace from 2010 onwards, newspaper coverage on this topic was not at all extensive. To a certain extent this reflects external control over the media. Overall only 12 articles expressed a negative attitude towards big chemical projects (2005, 12; 2008, 3; 2009, 3). Although limited in number, hard news reports had already started to voice different views, which is indicative of the general tendency of environmental agendas in the wider social context. But, of course, the limited number of critical reports on these projects contrasts with the importance of investigative reports that are the main force for articulating critical and even opposite viewpoints on these issues.

Popular environmental agendas becoming included in the national press reflects the degree to which society's attention was focused on environmental problems. Despite the priority for economic growth which was still encouraged by Beijing, the whole society was expressing its deep concerns over China's environment as a result of the emergence of environmental problems and health problems caused by the decisions made by Beijing over many decades. It is therefore no surprise to see that ordinary residents have learned that they have to protest, or, rather, are left with no choice but to protest against government decisions that would result in disastrous consequences for them, as mentioned in the previous chapters. Environmental problems have become an important trigger for social protests, which reveal intensive discussions among ordinary people. These protests represent part of the popular response to the authorities' determination to continue to boost the economy, demonstrating governments' indifference to people's reactions. For example, since 2012 angry people in Guangdong, Sichuan, Jiangsu and Zhejiang have marched in the streets asking local authorities to stop building industrial plants and other associated construction. Fierce clashes between civil protesters and local authorities have taken place. Protests of this kind are an indication of a radical shift in the general environmental discourse. The shift from the previous priority placed on economic growth to the struggle by civil society for a clean living environment represents the collective fear prevailing in society. Such a fear coexists with and resists the blind optimistic Promethean thinking at the top.

Meanwhile, the two discourses embodying Marxist environmentalism are not alone but resonate with currents in Chinese society that have been discussed in the Introduction. There have been diverse domestic voices about China's keen pursuit of the free market and modernisation. Prominent among these is neo-leftist thought. Scholars who are labelled as 'new leftists', Wang Hui for instance (Wang 2006, 2005, 2003), take a critical stance towards China's market reforms and neoliberal "capitalism with Chinese characteristics" and show sympathy towards the lower classes, for example the working class, who are seen as being sacrificed to economic reforms. They are quite sceptical about the impacts of capitalism on China's future. The two discourses we have identified, extinctionism and eco-equalism, are part of a wider public discourse of modernisation that holds a critical attitude towards capitalism and has grown out of the social context of China, echoing radical intellectuals' opposition to the capitalist production model. Going further than the neo-leftists, these two discourses also oppose the governments' collaboration with capitalism and the domination of elites as a result of the authoritarian rule of the regime.

The two discourses offer a Marxist environmental approach to examining modernisation. The critique of capitalism and the inescapable reality of suffering give economic modernisation a critical scrutiny and suggest ecological modernisation is impossible given the nature of capitalism. This obviously stands in direct opposition to what has been promoted by the Chinese government since the establishment of the PRC. Therefore, discourse of this kind embodies emancipatory potential as it speaks with an alternative voice.

The discourses of environmental problems have fully articulated environmental risk as a result of the speedy economic modernisation on which Chinese governments place such emphasis. They reflect on the whole modernisation development model of China. In these discourses, modernisation may result in a problematic rather than ostensibly prosperous life for ordinary people. It represents a loss of faith in progress and in the ability of the government to deal with the environmental problems emerging in the process. This is a discourse that prepares for reflexive modernisation in Beck's terms. In his terms a radicalised modernisation in a risk society has the capability of dealing with risks produced by "wealth production" by a means of knowing, preventing and minimising such risks (Beck 1992). It is important to raise the awareness of the public (including governments and policy-makers) of these risks caused by human economic activities. In a risk society such awareness would contribute to shaping relevant policy-making and implementation to deal with risks caused by modernisation. Following the Foucauldian perspective, the power exercised through the two discourses is subtle and capillary and makes individuals change their actions and attitudes towards the environment and modernisation as they are aware of the possible disastrous consequences they might suffer.

3.4 Conclusion

Drawing from framing analysis and discourse analysis, this chapter has discussed the discourse of environmental problems in investigative reports covered by ten newspapers from 2008–2011. The discussions in the chapter reveal the prominence of Marxist environmentalism embodied in the two discourses of environmental problems and their meaning for Chinese society and modernisation. The two discourses that have been identified in the analysis suggest a rival position that is against rather than for modernisation and economic development in China. They contribute to the shaping of public discourse of the environment and modernisation. Investigative reports are therefore playing a role in mediating environmental risks for their readers. This contributes to raising the public's awareness of environmental risks that can act as a catalyst for social change in many modern societies, especially a risk society (Beck 1992).

To understand discourses of this kind requires us to take a close look into the journalists' work that lies behind these reports. Chapter 4 will focus on analysing investigative journalists' work and their understanding of environmental problems in order to discover the tactics used by investigative journalists in negotiating the different parties who are aiming to influence their work, as well as to understand the importance of their personal judgement of environment problems to their reporting.

4
Environmental Investigative Journalists and Their Work

The previous chapter has discussed in detail the main features of the two environmental risk discourses identified in the coverage of investigative reports. A close link among environmental problems, economic activities and the failure of governance as well as between social and environmental inequalities characterises the two discourses, embodies prominent Marxist environmentalism and constructs an antagonism against state capitalism, the capitalist production mode and modernisation. Accounting for these features is how environmental investigative reporters report on environmental problems, which is the focus of this chapter.

From the social-construction perspective, news is far from being a mirror of reality (Schudson 2003). Reality constructed in news is only symbolic reality, which has no equivalence to objective reality, while our knowledge of the world is likely to come from symbolic reality rather than objective reality (Berger and Luckmann 1967). Thus, journalism is an important agent that mediates reality for us. Journalists function as bridges between our worldly knowledge and reality. Previous studies on the agent role of journalists (such as McNair 2003; Schudson 2003; Meyers 2007; Spyridou, Matsiola et al. 2013; Tapsell and Eidenfalk 2013) suggest that journalism constructs and shapes public discourse of reality that has impacts on our perception of the world and therefore leads to changes in society. Discourse constructed by journalism on certain issues influences, if not determines, whether and the extent to which the issues are understood, accepted or rejected by individuals and social groups. Social changes or changes in at least some areas might arise in the process. There are numerous factors that influence the nature of the discourse constructed by journalism. One important factor lies in the process of discourse construction itself, that is, how journalists construct the discourse of reality.

Social-constructionist scholars (such as McNair 2006; Tuchman 1978; Galtung and Ruge 1965; Cohen and Young 1973; Cottle 2006; Hansen 1991; Harcup and O'Neill 2001; Molotch and Lester 1975) generally view journalists as making news in a process of selecting, structuring, prioritising and defining events according to news values, editorial policies, organisational routines, personal schema and so on. Therefore, news values, editorial policies, routines and practices and personal schema are important factors influencing journalists' construction of reality. At the individual level there is a shared view among scholars that journalists' pre-existing frames and schema have an impact on journalists' news selection and production. However, little research has been undertaken on whether and the extent to which relevant personal experience, cognition and knowledge influences journalists' news production.

Investigative journalism is different from daily reporting. If compared with daily reporting, investigative journalism certainly presents some unique traits. Investigative journalism is not usually fed by official press releases and instead needs to collect wide evidence beyond official information feeds. News sources that speak in the coverage of investigative reports are outside the circle of elite news sources which daily reporters can access frequently and conveniently. It is entirely up to individual investigative journalists to decide who to talk to and whose words to include in reports. There might be less influence from organisational routines and structures in investigative report production and more flexibility at the individual level for investigative journalists' practices than in daily reporting. The initiatives of individual journalists appear to be very important in their practice. This is partly because investigative journalists have no deadlines to meet and enjoy more resources, such as time and money, to carry out investigations. They are able to spend extensive time doing research, such as visiting the scenes where events take place. This is in part because events or problems covered in investigative reports usually are not black and white, but often complicated and multifaceted. There is an individualisation tendency in investigative reporting. The understanding of the events and problems being investigated is to be gained through individual journalists' investigation, research and thinking which might vary from journalist to journalist. The nature of the research that investigative journalists conduct may determine the picture of the events they finally draw on in their reports. It is down to individual journalists to depict a full picture of the events out of the evidence they collect.

Nevertheless, how investigative journalists' own research, their presence and witness on the ground, personal experiences and feelings developed and gained in the process contribute to the construction of the meaning of events in their reports is under-researched. When journalists are making investigations, they somehow participate in, rather than simply observe, events and inevitably they develop their own feelings and judgement about them. How does that influence the role of investigative journalists? In the case of environmental investigative journalists in particular, environmental journalism is historically thought of as moving away from being objective journalism towards advocacy journalism (Wyss 2008; Schwartz 2006; Neuzil 2008). Are such journalists neutral observers or advocates? In addition, environmental investigative reports inevitably involve scientific knowledge; another important factor regarding the role of environmental investigative journalists is how they mediate between scientifically abstruse environmental knowledge and the public.

This chapter offers an interpretive analysis of the process through which Chinese investigative journalists report on environmental problems, offering explanations for the features of the discourses discussed in Chapter 3. The analysis emphasises the important role played by individual journalists in shaping the discourse of environmental risks in the investigative report production process – particularly how their personal experiences, (pre-existing) cognition, feelings and knowledge play their respective parts. In addition, struggles over discourse among different centres of power are manifested in the investigative reporting process, while individual journalists have demonstrated both tactics and strategies in dealing with struggles of this kind. It is the initiative of environmental investigative journalists to verify and contrast the words of news sources, including those of authoritative experts and officials, which underpins the two discourses of Marxist environmentalism identified in the coverage of environmental investigative reports. Because of the scepticism of investigative journalists towards the views of experts and officials and the contrast between the comments and information given by experts and officials, the so-called "primary definers" of environmental problems do not exist. The inclusion of voices of individuals who suffer from the "side effects" of a risk society, such as illnesses caused by environmental contamination, investigative journalists' personal experiences and feelings, and the description of the on-site scenes witnessed by the journalists themselves increases the level of cultural authority in their reports. The importance of investigative journalists' cognition of environmental problems as well as the existence of the fixed patterns in their choice of angles and frameworks for their reports also leads to the environmental discourses discussed in Chapter 3.

4.1   The profile of investigative journalists in this study

By carefully reading the environmental investigative reports published in the news media over the past two decades, we can easily recognise two types of authors. First, some media outlets have devoted designated pages or columns to the topic of environmental problems and assigned particular investigative journalists to regularly report on environmental problems and issues, such as Zhang Ke at *First Financial and Economic News*, Yang Chuanmin and Yang Xiaohong at *Southern Metropolitan Daily*, Lv Minghe, Lv Zhongshu and Bao Xiaodong at *Southern Weekend* and Gong Jing at *Caijing* magazine at the time when the author did the main fieldwork in 2011. Usually these media outlets have already institutionalised and integrated environmental investigative journalism into their daily organisational practice.

Second, in other media outlets, journalists occasionally investigate and report on environmental problems. They produce reports on the topic for these outlets in a random way; neither investigative journalism nor environmental reporting has become a routine practice. These investigative reports are usually carried at a time when environmental incidents occur. To take an example: chromium pollution took place in Qujing, Yunnan Province, in 2011. After the Xinhua News Agency exposed this incident, a considerable number of news media across China reported on it. As a result, there was a rapid increase in the number of environmental investigative reports published in that year. However, for most of these news media, such as *Information Times* (*xinxishibao*), the reports on this event and the topic of environmental problems generally are only one-off, lacking duration. The investigative journalists who reported on this incident are not responsible for reporting on the topic of environmental problems in their daily work. This study focuses on the first type of environmental investigative journalists.

In total, 42 environmental investigative journalists were interviewed for this study from 2011–2013. They were selected for interviews largely because they have constantly published nationally influential environmental investigative reports over recent years. Their reports are among the most important investigative reports on environmental problems recently published. For example, China Dialogue, an NGO launched in 2006 in China, give annual awards for "The Best Environmental Journalist" and "The Best Environmental Report" in collaboration with the *Guardian* in the United Kingdom, with the aim of promoting the common understanding of environmental problems in China. Chinese environmental journalists see these two awards as a great honour (Interviews, 2011–2013). Some of the interviewees have already received these awards in different years for their excellent journalistic practice and reports.

The majority of these investigative journalists are male (less than 10 out of the 42 journalists are female). All of them are employed full-time by influential news media that favour and support investigative journalism, such as *Southern Weekend*, *Southern Metropolitan Daily*, *First Financial and Economic News*, *Caijing* magazine, *New Century* magazine and *China News Weekly* magazine (*zhongguo xinwen zhoukan*). Most of them are university-educated and some of them have postgraduate degrees. However, except for two of them who studied physics in university, the majority of these investigative journalists have degrees in literature, history, journalism or other social science and humanities subjects. They are a group of experienced journalists who have at least five years' working experience as journalists and more than half of them have ten years or more experience working in news organisations. Most of them are in their late twenties or in their thirties.

This group of journalists developed their interests in environment-related topics at different stages of their careers. Some of them began to report on environmental problems immediately after they took on the occupation of journalism, but some of them had worked as investigative journalists for a quite long time before they embarked on this area of reporting. Very few of them were actually assigned to report on environmental problems by their superiors. Most of them started reporting on environmental problems as a result of their own interests. According to them, generally speaking there are four main reasons that account for their interest in environmental issues.

First, investigative journalists pay attention to environmental issues because, in the words of the interviewees, this is not only that "environmental problems have become an important aspect of Chinese society" but also because "the public pay much attention to environmental problems, especially when relevant problems become increasingly pressing and closely related to everyone in society" (Interviews, 2011). The topic of environmental problems thus is judged to have high news values. Second, investigative journalists embarked on environmental topics in order to look for chances and changes to develop their career paths. This is part of their career plan. Lao Xi, to take an example, shifted from reporting on social and political issues to environmental issues after he had worked as an investigative journalist for more than ten years at an influential metropolitan daily. Third, some of these investigative journalists suffer or witness environmental problems in their own living space. An interesting example that emerges across the interviews is the one of protests against rubbish-burning projects in Guangzhou. For those interviewees who have reported on this event, one major reason to engage themselves in the reporting of these protests as well as the relevant campaign is that they live in the residential districts that would be directly affected if the projects were to be launched there.

The fourth reason is exemplified in the view of Qian Qiang that environmental problems are more important than many other social and economic problems because they raise questions about human beings' fundamental existence. Qian Qiang explained how and why he started focusing on the topic of environmental problems in the following words: "in 2003, I was still working at the XX.[1] An environmental news studio under the Environmental Protection Ministry invited me to attend a press conference hosted by the Environmental Protection Ministry. This was the first time I was involved in this area. Later on, I did a report on environmental problems that was successful and influential. ... After I joined the YY,[2] I initiated the wish to work in the field of environmental protection. ... Because I think economic issues, such as issues about enterprises, are only concerned with the here and now and with making money, and about fights between the general managers or the CEO. But in my view, in the long term ecological and environmental problems are about the fate and future of human beings. Climate changes, including natural disasters, may destroy everything you are working on in relation to the economy. ... This [environmental problems] is an issue that is long-term and in human beings' vital interests" (Interview, 2011).

From the accounts these journalists have given to explain why they have started investigating and reporting on environmental problems, one can conclude that environmental problems have emerged to be one of the most important aspects of Chinese society. News organisations and journalists regard them as having high newsworthiness and that they are likely to capture intense public attention. Topics of this kind are also intimately linked to the direct life experience of journalists, who may be environmental victims themselves. This point also reflects the innate uniqueness of the topic of environmental problems. While social, political and economic issues may be significant, journalists might well be outsiders to the issues they are reporting. This is not only because of the social class they belong to, but also because of the circumstances in which they live. However, the topics of environmental problems are indeed seen as class-free and involve everyone who lives in that environment. Journalists themselves can never be neutral observers and outsiders in relation to environmental problems, since they are involved to a certain extent.

4.2 Environmental problems in the mind of investigative journalists

Environmental investigative journalists interviewed for this study appear not to be descendants of Prometheus who are willing to convince themselves and the people that the existence of fire can save the earth. William L. Lawrence, a famous American science journalist, once described science writers as "true descendants of Prometheus" who "take the fire from the scientific Olympus, the laboratories and universities, and bring it down to the people" (Weart 1988: 99–100). In the Chinese context, as discussed in Chapter 1, the Promethean view believes in man's ability to conquer nature and in technological progress to improve economic development and to fix associated problems. Obviously, environmental investigative journalists do not share this view.

Environmental investigative journalists hold clear and prominent perceptions of the environment and its problems. To a great degree they share common views in regard to how they view the relationship between humanity and nature as well as the categories of environmental problems. When asked to comment on these two questions: (1) what is the relationship between human beings and nature, and (2) what kinds of environmental problems exist in China in your eyes, journalists presented two types of discourse that reflect their views of the environment and its problems. The first discourse is the doom discourse. Journalists generally see the development of human civilisation as standing in opposition to the harmony of nature, with some nuanced different views. There are two levels of meaning in their interpretation of the humanity–nature relationship. On the first level they tend to see the unsustainable activities of humanity as destroying the harmony between human beings and nature. Nature is seen as providing resources for human development. However, resources are limited. If human demand for resources does not consider the ability of nature to provide, human development will inevitably destroy nature. For example, most environmental investigative journalists see pollution, the construction of dams across the country and desertification as the three most severe environmental problems that China is facing. These problems are a result of extreme economic activities in China. Human activities are responsible for environmental problems and even for disasters such as landslides and droughts.

On the second level they tend to think human activities are negligible and petty if compared with the power of nature shown in, for instance, natural disasters and desertification. The kind of impairment caused by human activities is unable to destroy nature completely, since nature is much more paramount and permanent than humanity. Despite that, this kind of damage will inevitably devastate the environment that is suitable for humans to live in and therefore eventually annihilate humanity itself. This interesting argument, shared by a number of these environmental investigative journalists, is exemplified in the following summarised comments of Yang Zheng and Long Guo: humanity's history is extremely short, if compared to that of the earth; and human activities are powerless, if compared to the power of nature. If humanity disappeared from the earth, nature would recover to its original status within a very short period of time. What humans destroy is not nature, but humanity itself (Interviews, 2011). As such, this is a kind of doom discourse that sees human activities as unavoidably causing damage to nature and such damage in turn will result in the downfall of humanity.

The second discourse is a discourse of diffidence towards the capability of governments to deal with environmental problems and issues. Rather than showing confidence in governmental ability, almost all environmental investigative journalists regard governments as having exacerbated or even caused environmental problems. More specifically, most journalists blame the "system" (*zhidu* or *tizhi*) – governance in general and particularly environmental governance – for the destruction of nature, and believe that the problems with the "system" have been more prominent than the capability of new technologies to fix up environmental problems. The following remarks of Lin Qun are representative of this view: "I feel the most severe environmental problem we are facing nowadays in fact is the problem of environmental governance. By saying environmental governance, I do not merely refer to the environmental protection ministry, but the problems existing in China's whole (political and social) system. Let me explain this for you. Soil pollution or problems about underground water are good examples. Even if one day we would eventually find solutions through the use of advanced technologies, the flaws embedded in the current [political and social] system itself, which I think perhaps are the biggest problem, make it [the fixing of the problems] impossible. This has been changing over recent years but actually the changes are very tiny. So behind all these [environmental] problems is the problem with and in the system" (Interview, 2011).

Apart from that, part of this lack of confidence comes from journalists' worries over the tremendous pace of economic growth and the engagement of governments in frenzied pursuit of GDP, which leads to the intimate collaboration between governments and enterprises as well as the reality that local governments even use environmental protection as a good reason to request financial subsidies. When asked what images come to his mind if he starts to think about environmental problems, Wang Qing, an investigative journalist at *Yunnan Information*, said his first response was the Dianchi Lake (a local lake in the Yunnan Province). He then went on to explain why: "the local government spent several thousand millions of RMB to clean the Dianchi Lake but all efforts failed. Then the local government used the Dianchi Lake as an excuse to ask for more money from governments at higher administrative levels. For this one lake, the local government opened up and kept a department at the administrative division level with so many staff working for the department. Every single year, the construction and cleaning of the lake is repeated over and over again. This example, I feel, can really illustrate the whole situation of the environment in China." After making the comments, he then continued to explain the ridiculous way in which local governments' environmental pollution management has contributed to the growth in local GDP. With this discourse, environmental investigative journalists show a strong scepticism towards the incompetence of governments and their poor environmental governance.

These environmental investigative journalists have enumerated the most severe environmental problems China is currently facing and their causes, as discussed in detail below. Environmental problems identified and mentioned by these journalists can be summarised and classified into seven categories. More importantly, from the perspective of investigative journalists, all of these environmental problems are problems with and in China's economic, political and social systems rather than problems about nature. That is to say, environmental problems are social problems, political problems, problems with the system.

In addition, one interesting point that is coherent with the analysis of the agendas and discourses of environmental reports in the previous chapters is that journalists seldom mentioned climate change or global warming when they were asked the question: "What are the biggest environmental problems in China in your view?" The comments of Guang Ning illustrate well the reason why journalists exclude these internationally important environmental problems: "Our current biggest agenda for environmental problems would never be climate change or low-carbon consumption. The so-called global warming and low-carbon consumption are only the fashionable code the elites use. We say this kind of thing only because we would like to show we are sophisticated. The ordinary people would not care much about these things as they are too far away from their real lives, when they are drinking toxic water and eating poisonous food" (Interview, 2011).

Pollution

Pollution is the number one item on the list of the biggest environmental problems identified by journalists. According to them, almost all basic elements – especially water, air, soil – in China now suffer pollution. Water, air and soil pollution are the three primary pollution problems. Associated with these pollution problems are health and food security risks, such as rice and vegetables contaminated by heavy metals, cancer villages and high lead-poisoning levels. These pollution problems are mainly caused by industrial activities, such as manufacturing and mining, and factories, such as the operation of chemical factories or even PX projects and of rubbish-burning projects.

Water crisis

A considerable number of interviewees regard the water crisis as a very prominent environmental problem in China. Apart from pollution, the water crisis generally includes two other types of problem: water shortages and the changes in the water system and ecology in rivers, lakes and underground water. The latter is further seen as the cause for the former. Human activities, such as the construction of numerous dams and hydroelectricity stations, are seen as the primary interference with the water system in nature. The construction of the Three Gorges Dam is one of the major human-induced threats to the water system appearing in the discourse of journalists in these interviews. Almost all the journalists showed their doubts about the construction of the Three Gorges Dam, though to different degrees. Some of them even referred to it as a disastrous project, since the building of Three Gorges Dam would inexorably change the natural structure of the earth and the natural environment of underground water, rivers and lakes. Although they admitted that it would be difficult to prove that there are direct connections between the construction of Three Gorges Dam and natural disasters in recent years, they thought the problems caused by the dam to the environment had already emerged. In addition, the rapid urbanisation in China, especially in Beijing and Shanghai, has dramatically increased these metropolitan cities' demands for water. This is also seen as causing the two types of water problems. Several journalists also mentioned that the underground of the whole Huabei (Northern China) plateau has already become a huge and empty funnel, because underground water has been used up.

Desertification

Desertification is also on some journalists' lists. Desertification includes that of grasslands and of forests and mountains. Journalists have even identified connections between desertification and natural disasters, such as landslides, as a result of the evidence they have collected in investigations. For example, several investigative journalists who have repeatedly reported on landslides hold this view. According to them, their investigations show that the destruction of local forests and vegetation on land and mountains usually exacerbates landslides in places such as Zhouqu. The destruction of forests and vegetation is usually caused by the rush for economic growth and associated activities, such as over-logging.

Hollow ground and resource shortage

Hollow ground is usually caused by over-mining in those areas that were proud of their enormous natural resources in the past. In the words of investigative journalists, this environmental problem thus is often linked to the problem of resource shortages that is a consequence of unsustainable development. Some journalists have identified places such as Shanxi as suffering from this problem and explained the reasons that resources in those places have run out. Resources are what politically and economically advantaged individuals and interested groups endeavour to seize and exploit; they are the basis for economic growth and one of the fundamental causes of environmental problems.

Destroyed ecology

Species have disappeared or are disappearing from China. Their disappearance in some cases means some species have moved to other places that are more suitable for living, but in other cases means extinction. Their disappearance is a result of the destruction of ecology in seas and on land. According to investigative journalists interviewed in this study, the construction of projects such as dams and artificial islands offshore, and other environmental problems caused by human activities, such as pollution, should be held responsible for that. Porpoises, to take an example, became extinct as a result of food shortages, pollution and human activities, such as dredging rivers and lakes, and dams and shipping which stopped their seasonal migration. Soil salinization and desertification are two examples of destroyed ecology that come about as a result of human activities and lead to the reduction of arable land and an increase in the incidence of natural disasters.

Natural disasters

Some investigative journalists classify natural disasters such as landslides and droughts as environmental problems. More specifically, natural disasters are seen as being caused by environmental problems. The majority of journalists regard these as to a certain degree human-induced rather than natural disasters. Two natural disasters that are often mentioned by these journalists are the landslide in Zhouqu in 2010 and the severe drought in the area of the Yangtze River in recent years. The Zhouqu landslide was deemed to have been exacerbated by desertification caused by over-logging, while the droughts in the region of the Yangtze River are thought of as related to the construction of the Three Gorges Dam and its reservoirs.

As such, we find there is an interesting match between the way environmental problems are defined and conceived in the minds of investigative journalists and the coverage of investigative reports. This manifests itself both in the types of environmental problems covered, and also in the interpretation of them as social problems and in the connections drawn between environmental problems, economic activities and governance. This match also exists between the representation of social and environmental inequalities in the reports and the link between the exploitation of resources by the privileged and the environmental costs and consequences paid and suffered by the unprivileged. The coherence between the media and mind frames explains well why investigative reports frame and interpret environmental problems in the way that has been demonstrated in the previous chapters.

4.3 Investigative journalists' perceptions of environmental journalism and their roles

Investigative journalists remarkably share a view that sees environmental investigative journalism as not differing much from investigative journalism on other topics in terms of reporting techniques, practices and news values. Like investigative journalism in general, environmental journalism involves long-term, extensive investigations and the revelation of problems in society, requiring the practice of balanced reporting. The average cost required for completing an environmental investigative report was from 7,000 to 8,000 RMB in 2011, which was similar to the average cost that investigative reports on other topics needed at that time (Interview, 2011).

There are three prominent features that distinguish environmental investigative journalism from investigative journalism on other topics. First, the successful practice of environmental investigative journalism requires journalists to be more professional in the sense that it requires them to have more professional knowledge than investigative journalism on other topics. Here, the term "professional" has a double meaning: to be professionally objective and to have professional "specialist knowledge" in the areas of their reporting expertise. By "specialist knowledge" they mean scientific knowledge about relevant issues surrounding certain environmental problems, such as pollution, energy, geology, vegetation and so on. They tend to see themselves as "expert journalists" or "academic journalists" who have developed expertise in one or more areas in relation to environmental problems and are adept at carrying out research. In addition, compared to investigative journalism on other topics, environmental investigative journalism has a closer collaborative relationship with NGOs and scientists. Investigative journalists express their willingness to collaborate closely with NGOs and participate or even initiate NGO activities on some occasions, although they refuse to see themselves as activists or advocates instead of professional journalists.

Second, while reporting on environmental problems, investigative journalists are able to observe events before they start interviewing relevant news actors, so that they can collect empirical evidence through fieldwork about what has happened, its causes and consequences, prior to carrying out interviews, whereas investigative journalists reporting on social and political issues, especially on political scandals and wrongdoing, lack this kind of advantage. In their terms, "the evidence is out there for you to witness with your own eyes" (Interview, 2011). As a result, this advantage reduces the level of difficulty in investigations and is an important reason for the proliferation of environmental investigative reports in recent years. It is thus natural for these investigative journalists to see the ability and willingness to carry out research as an essential part of being environmental investigative journalists. They regard themselves as being good at conducting fieldwork and making judgements about events based on evidence collected by using all kinds of fieldwork methods. Evidence-based judgement-making is deemed to be central to the work of environmental investigative journalists. Judgements made about particular events usually offer frameworks for their reports on them.

The third feature, closely related to the second point, refers to the low level of difficulty in carrying out investigations. Apart from ease of evidence collection, the lower political sensitivity of the subject matter also contributes to more convenient investigation. For example, when asked in what ways environmental investigative journalism differs from normal investigative journalism, Hao Gang replied that environmental investigation is easier than investigation of other topics. This is because "other investigations, such as investigating official corruption, are necessarily politically sensitive. Nevertheless, now [for environmental investigative journalism] even the ruling Communist Party promotes the idea of sustainable development and other concepts, such as clean production and harmony between humanity and nature and so forth. ... Therefore, they [governments and officials] do not completely exclude you from reporting on relevant topics" (Interview, 2011).

Wang Qing, an investigative journalist at *Yunnan Information*, summarised the features of environmental investigative journalism well in the following words: "investigative reports are usually related to muckraking and public opinion supervision and the political authorities. They usually talk about something hidden from the public's eyes and something the political authorities try to cover up. However, environmental investigative reports must be related to environmental protection and require journalists to have professional quality, professional knowledge and professional skills in order successfully to complete the reports. Journalists should understand the industry [field] very well. In addition, environmental investigative reports have more space and topics to report on than other topics for investigation. That is because environmental investigative reports can be about environmental pollution, can be about social renovation in the field of environmental protection and can be about people and society."

According to these journalists, the topics that environmental investigative journalists usually report on refer to problems with the environment (matching the types of environmental problems in China that they have identified), emergency events and incidents that will generate enormous impacts on the environment or disasters that imply severe environmental problems. Very few journalists regard topics such as clean technologies, clean energy, chemical fertilisers, global warming and climate change as mainstream topics for environmental investigative journalism in China. This, of course, is attributable to their own understanding of environmental problems.

4.4 Initiatives by investigative journalists to verify and compare information

Environmental investigative journalists have shown strong initiative in order to verify and compare information received from various news sources, ranging from experts and officials to environmental victims. This initiative enables their reports to encompass the possibilities of individual interpretation of environmental problems as influenced by their personal cognition of these problems. In contrast with daily reporting, environmental investigative journalists have enough time to collect, digest and verify information before they package it together for their readers. This process liberates investigative journalists from reliance on official news sources and news beats. A variety of information is collected through methods which include interviews with experts and ordinary people and their own research, observation and analysis. Materials involved range from interview recordings, conversations, academic reports and papers to what journalists have witnessed on the ground and relevant personal experiences and feelings. In the process of verifying information and making judgements, investigative journalists work like researchers and anthropologists and adopt research skills. They treasure this kind of research, regarding it as critical for good practice. They need professional specialist knowledge of relevant environmental issues, knowledge and cognition about certain environmental issues gained from their direct personal experiences in the past or at the present time, and their analysis of the whole situation. The verification process takes place in an escalating fashion by which journalists analyse all the information they possess and draw conclusions about what has happened out there in the environment.

Journalists gain professional specialist knowledge in two main ways. They usually turn to scientists and other experts, such as NGO staff and officials, who have expertise in the relevant areas. They interview or talk to experts to get information and views on certain topics in order to gain an initial and overall understanding of specific environmental areas or gather different or even opposing viewpoints regarding environmental topics for their reports. However, journalists see what experts provide as merely original raw materials that they may or may not use in their reports. On some occasions they may doubt the motivations of experts – although not all experts – in making certain comments and perceive some connections between experts' comments and their interests. Journalists can be sceptical of the level of honesty of some experts' answers. According to investigative journalists, the standpoint of experts may be affected by their interests and connections to the ruling Party. Some experts refuse to speak what is really in their minds, partly because they do not dare to do so but partly because they belong to specific interest groups. For example, Gao Qing condemned some hydroelectric experts for supporting the governments' construction of hydroelectric projects for their own interests. When talking of one expert's view that food additives are not associated with food security problems, Wang Dong expressed views that differed from those of the expert and believes journalists should raise questions and doubts in order to represent the food additive issue accurately in their reports. Journalists are aware of the intimate and covert ties between experts and the authorities. For instance, Sha Ming used the example of Huang Wanli to explain the importance of such a relationship. Huang Wanli is an expert who has been isolated by decision-makers because he has publicly opposed the construction of the Three Gorges Dam. This is an example that illustrates the point that experts can choose to oppose policymakers' decisions; but the consequences may be that they are cut off from the decision-making process and exiled from all kinds of official forums and debates.

As such, for these investigative journalists, it is important not to be "led or guided by experts". Instead they have to remain independent in making judgements about specific things. The selection of experts' views or quotes is guided by journalists' own judgement and standpoint. In other words, journalists like to retain journalistic authority and merely treat experts as one type of news source, who should not be the "primary definers" of reality. In order not to be "led or guided by experts", investigative journalists often highlight the importance of conducting their own research. Their expertise in specific environmental issues has developed from their extensive research, such as reading academic papers and research reports. All investigative journalists have expressed the view that they work like researchers who need to read widely in the areas of their reporting topics. This helps journalists develop "professional specialist knowledge" in these specific areas so that they can be in a good position to communicate with experts and officials and to understand their views and comments. On the other hand, investigative journalists should gain the ability and knowledge they need to make the right judgement about the situations they investigate, especially when journalists have realised that some "experts are on the side of interest groups".

Another method journalists usually adopt to understand the situations involved and make judgements is observation. Journalists even believe that observation should precede contacting experts, especially officials. As discussed above, environmental problems are out there and can be witnessed with the investigative journalist's own eyes. It is the normal strategy for environmental investigative journalists to collect empirical – especially eye-witness – data before approaching officials. This is because approaching officials may mean inviting official interference into journalistic practice. In general, journalists regard fieldwork research as an essential part of practising environmental investigative journalism. Xi Bei, an experienced environmental investigative journalist who has expertise in desertification, for example, believes that environmental investigative journalists should go into the field, as this is the best way to understand events (Interview, 2011). When reporting on the big crisis facing offshore ecology, Li Jing spent nearly 20 days walking through all these offshore places from south to north, until he finally found out what has happened to offshore ecology. In this sense, environmental investigative journalists work like anthropologists who go into the field, collect empirical data and sometimes even participate in field activities. Such experiential knowledge also involves journalists' interviews with ordinary people – especially environmental victims – and witnessing the daily lives of environmental victims, which results in journalists being able to include victims' voices speaking of the effects have suffered from environmental problems, and increases the cultural authority of journalists' reports.

In journalists' information-verifying process, cognition plays an important role. Cognition refers to the schema in journalists' minds that has been developed over years based on their life experiences. For example, journalists' cognition of environmental problems and understanding of the relationship between humanity and nature have been demonstrated to have impacts on their knowledge of the specific environmental issues that they are reporting on. Several journalists who have reported on pollution and water problems drew on their life experiences in their hometown to understand the situation and to make judgements about these environmental problems. To take an example: Li Du grew up in midland China and saw the rivers there every day when he was a child. When he reported on the severe water problems in midland China (both drought and flooding and even dam collapses), his life experience told him that the water problems in recent years have some relationship to the building of dams that are used to reserve a huge amount of water in an unnatural way that has changed the underground water ecology between rivers and lakes in the region (Interview, 2011).

Journalists' understanding of the research-based nature of their work, nevertheless, has no impact on their journalistic identity. Although they stress the importance of participation, observation and research, investigative journalists interviewed in this study still see themselves as journalists rather than witnesses on the ground, researchers, experts or activists. When asked about their identity when they participate in NGO activities or carry out fieldwork, they reply "I am a journalist for sure" without any doubts. Their apparent occupational consciousness as journalists and their interpretation of their work, however, conveys a sense of subjective, participatory and advocacy journalism.

4.5 Judgement, the choices of angles and routines

After journalists have collected information from all kinds of sources and feel these data are enough to enable them to arrive at a conclusion, they start to examine all the evidence, try to digest the information properly, make a judgement and develop a framework for writing up their reports. Even in the middle of their investigation, they start making judgements and forming a framework based on the evidence they have collected. Just as Li Jing said: "it [to make a judgement] is impossible at the start of the investigation. You have to gain a basic understanding of the issue you are investigating, before you can make a judgement about what it is and decide in which direction you should work." In their terms this is the "logic" (*luoji*) that needs to be developed out of a chain of evidence (*zhenjulian*). For these environmental investigative journalists, judgements cannot be made in the absence of this logic, or framework, which is crucial to their reports. This is for two reasons. First, journalists need to practice balanced reporting, that is to say, they need to represent different opinions in their reports. However, balanced reporting does not mean an absolute lack of judgement and merely the presentation and listing of all opinions and information. Reports would in this case become dialogic texts and be essentially meaningless. Second, journalists need to remain independent and avoid being guided by one or two news sources; at the same time they also need to guide their readers in how to understand the issues they are talking about in their reports. Therefore this kind of logic or framework derived from a chain of evidence is necessary for reporting on environmental issues. This of course indicates the independence of journalists but also reflects the biases and meanings embodied in their reports.

The necessity of making judgements and choosing angles for their reports characterises investigative journalists' view of their job. The judgement, choice of angles and potentially emergent routines explain the features of the frames of the investigative reports analysed in the previous chapter. The work of investigative journalists occurs in a process whereby they develop knowledge and make judgements about the subjects on which they are reporting based on the evidence and information they have collected, and then choose the angles of reporting to present the subjects. Making judgements or the choice of angles is not something taught by journalism textbooks or training. Instead, these are learned by doing, from their practices. These ways to make judgements and choose angles inevitably lead to routines.

The judgement of the frameworks of investigative reports starts from their evaluation of the topic. For the investigative journalists interviewed in this study, a principal criterion of newsworthiness is whether a topic is seen to be close to the interest of the public and also in the public interest, which gives it social value. This criterion decides whether a topic comes to their attention and deserves their attention. For example, when Ming Huang explained why she went to report on desertification, she said: "I heard someone talking about this when I was travelling in that area. I found the topic interesting so I did some research and decided this was a topic of interest to the public and thus is worthy of my attention." She then explained it to her newsroom and the latter approved her idea because "this topic is meaningful. It has social value as it is relevant to the public interest". The understanding of what is the public interest thus becomes most interesting if one wants to understand what can be classified as having news value for investigative reports. There is a shared view among interviewees that events fitting the category of in the public interest are usually those involving a large group of people, reflecting the fundamental problems in Chinese society, or having enormous consequences to society. The usual understanding of public interest may, of course, result in routines in making judgments and fixed patterns in the frames and interpretations of environmental problems in their reports.

Certainly the public interest is not the only criterion on which judgements are made about the newsworthiness of the topic. Most items on the list of newsworthiness and news values as suggested by Galtung and Ruge (Galtung and Ruge 1965) for daily reporting are also applicable to environmental investigative reporting. For example, despite having social value, a topic will be viewed as mundane when it has been reported over and over again without any uniqueness or exclusiveness. As regards the topic of cancer villages, for example, as Guang Ning explains, "investigative journalists have repeatedly revealed the problem of pollution and its association with the emergence of cancer villages. However, most of the reports have had no effect. This makes the topic less and less attractive for news media and investigative journalists" (Interview, 2011).

However, unlike investigative reporting on other topics, environmental investigative reporting usually does not involve the revelation of hidden scandals that represent a big scoop and an exclusive story. Except for a few topics such as environmental incidents and organisations' and individuals' wrongdoings, for example, in relation to dam and artificial land construction, most of the topics concerning environmental problems are not "new news" but "old news". That means these topics, such as cancer villages, desertification and resource exhaustion, have already been reported widely and talked about for a while in China. But investigative journalists still want to report something new out of these old topics and to make their reports exclusive. These investigative journalists recognise that making unique judgements on the topics and choosing unique angles are exactly what make their reports exclusive.

Making reports exclusive is conditional on the application of journalists' own judgement on the situation that underpins their report. Just as Lin Qing said, obtaining core information and making unique judgements are the two different premises for having an exclusive story. If you do not have exclusive core information, then you should underpin a unique judgement by choosing unique reporting angles. This is also a way of maintaining journalists' independence and professionalism, according to these journalists, although this process may lead to biases in their reports. Thus, looking for a unique angle is an effective way for these investigative journalists to generate exclusiveness. This is because different investigative journalists might choose different angles that generate different interpretations of the same stories.

One way to make investigative reports exclusive, for example, is to recover truth through locating the topics in social contexts. Ying Dan published a series of influential in-depth investigative reports on the Three Gorges project at *Liaowang* news magazine, under the Xinhua News Agency. This series of reports reconstructed the historical debates on the building of the Three Gorges Dam, going back to the establishment of the PRC. He regarded it as difficult to measure quantitatively the advantages and disadvantages of the project because the project had too many disadvantages and merits. What he was able to do was to "restore the process and reconstruct the debates [between opponents and supporters of the project]" (Interview, 2011).

However, choosing angles for reports is a result of investigative journalists' reporting habits, forged in their journalistic careers. The reporting habits result in the preferred analysis framework of investigative journalists. Which angle or which analysis framework investigative reports eventually present is closely related to the cognition of the journalists and the high-ranking editors at news media outlets. The following comments of Wang Dong, a *Southern Weekend* journalist who is famous for his consistent reporting on the construction of dams and hydroelectric projects, are representative of this point: "if a journalist who is good at reporting on social issues reports on that [the dam construction], s/he tends to seek and focus on conflicts. The Nujiang River dam is likely to be reported on from the perspective of social issues, with a stress on the conflicts between local residents and local governments and construction companies. For the topic of high levels of lead in the blood, if the report only highlights the occurrence of high levels of lead in the blood itself, it is a pure scientific and environmental report. If the report stresses the cause of these levels – land pollution – the topic is treated as a social environmental event. If the report focuses on the conflict between governments and victims, it is then treated as a social issue." He then went on to explain his preferable analysis framework: "the report [on high levels of lead in the blood] might involve where the main enterprises that are to blame are based, how they have been transferred, relocated and expanded to other places, what is the purpose and what is the reason [for the transferring and relocating]. Why it has not been stopped when government measures were strengthened. [My understanding is] when governments tried to deal with that, enterprises tried to move to other places where they are inviting investment, which is the living foundation for these enterprises. The demand in the downstream market is the reason for the existence of a large number of workshop-like manufacturing enterprises. I prefer to take this angle" (Interview, 2011).

The choice of angles may lead to moral judgements and a fixed or routine framework to interpret environmental problems. The dichotomy between governments and people, between the (political and social) system and the people or between economic development and the environment now has become a major analysis framework for investigative reports on environmental problems. This explains why the conflict frame has been found so obvious in the 258 environmental investigative reports. For example, Guang Ning, a former director of an investigative team at a metropolitan newspaper and the deputy editor-in-chief at a commercial news magazine in 2011, justified the dichotomy through making comments on the reporting on cancer villages: "China now sees the occurrence of insoluble conflicts every day. News media that have any sense of the public interest cannot avoid following this dichotomy. For example, for environmental investigative reports, the most common topic is the pollution caused by chemical factories. People in so many villages are dying, which is the worst thing in the world. However, the death of people cannot stop chemical factories from polluting the environment, even after being reported by the news media. This problem cannot be resolved at all. What can we do? The rift and conflicts are the prevailing discourse system that investigative reports can follow" (Interview, 2011). Another reason for this moral judgement and the dichotomy is that most environmental investigative reports are an extension of reports on social issues. "Most environmental investigative reporters used to report on social issues. Therefore they tend to take the analysis framework of seeking conflict and report from the perspective of moral judgement" (Wang Dong, interview 2011). This moral judgement investigative journalists make offers some useful explanations for the features of the discourse in investigative reports.

4.6 Difficulties in reporting and the guerrilla tactics of investigative journalists

As discussed already, these journalists tend to see environmental investigative reporting as suffering fewer reporting restrictions than investigative reporting on other topics, especially social, political and economic issues. This is because reporting on social issues would likely touch political nerves, such as corruption and wrongdoing. Nevertheless, reporting on environmental problems usually would not. The fact that governments and officials also want to have a clean environment is another reason that political authorities are more tolerant with environmental investigative journalism than investigative journalism on other issues. Besides, the chance for new markets and opportunities created by environmental investigative reporting makes it welcome by commercial organisations at some points.

The comments of Guang Ning on the topic of environmental problems are typical of this argument. During the time he was the director of the investigative reporting team at an influential commercial metropolitan newspaper in the early 21st century, investigative reports on environmental problems started flourishing in China. When commenting on when and why the topic of environmental problems became popular, he recalled: "environmental protection suddenly became one of few topics that we were allowed to report. At that time, I felt this topic was the only one which could break through propaganda control and on which we could make good news reports. Therefore, at that time, you may find investigative journalists on social issues all turned to reporting on environmental problems, because this topic was the only one that would not be spiked by high-ranking editors. That period of time was from about 2005 to 2007. Of course this does not mean there were no limitations. It only meant that compared to other reports, the reporting limitations were fewer and the chances of having post-publication troubles were relatively low. This was because at that time the central government and even the provincial governments wanted to control pollution and to clean the environment. At that time, some big ideas, such as "scientific perspective to development" (*kexue fazhan guan*), and "green GDP" were raised by top political leaders. The central government's policies were encouraging and generating room for this type of reporting. Politics often affect Chinese news media a lot. When there was some space for this topic, news media across the country took their chance and turned to reporting on this. I feel, during those years, Chinese environmental investigative reports developed rapidly" (Interview, 2011). His comments make several interesting points for understanding environmental journalism reporting: 1) there is a need by the central government; 2) the topic of environmental problems is relatively politically safe and thus has fewer reporting bans and prohibitions; 3) news media have a willingness to report on problems rather than stick to positive reporting.

Nevertheless, environmental problems often involve conflicts among multiple news actors with their respective interests. It is not difficult to identify conflicts between environmental problem generators and victims; the former would want to escape from being held accountable and the latter seek environmental justice. A closer look into China's social context and media–government relationship will reveal a complex picture in which local governments may not want environmental problems occurring in their administrative territories to be exposed for various reasons. Environmental problem generators are often enterprises that are under the protection of local governments. The two share common interests. In some cases environmental problems may even see the direct involvement of governments and local officials. In addition, environmental problem generators sometimes are difficult to identify and judge, since in some cases cross-territorial governments and enterprises are involved. NGOs interfere in the relationship among environmental generators, victims and governments. Taking these factors into account, it is not so much about whether journalistic practices in reporting on environmental issues are limited by bans or other reporting restrictions. It is more a question of how journalists deal with the interference of these news actors into their journalistic practices in order to retain their independent stance.

The basic strategy environmental investigative journalists adopt is to use "right methodologies". By "right methodologies", journalists mean using different methods to get different types of information. According to journalists, before starting an investigation, they have to know what information is needed for producing a successful report and where to get it. For example, Xiao Feng at *Caijing* magazine commented: "you should go to the right places and ask the right people for the right information.

If you feel you cannot obtain information from enterprises and governments, you should not rush to ask them to provide the information for you. You have to look for the information in your own ways. Of course after you have already obtained the information you want, you then go to governments and enterprises for some core and key information. Basically our methodology is like that" (Interview, 2011).

Research usually is the first step most environmental investigative journalists take to start their investigation. Many environmental problems involve academic and scientific knowledge that journalists may not be familiar with. Some of them consult experts of their acquaintance for an initial understanding of the subject while most of them tend to gain relevant knowledge through their own academic research, that is, searching and reading relevant academic reports and papers. For example, Shang Hua said she would read a lot of academic materials and set off for fieldwork investigation with doubts and questions she had derived from her reading before asking for information from officials.

As discussed already, observation and fieldwork is an essential part of journalistic work for environmental investigative journalists and also a strategy they all use. In contrast with other types of investigative reporting, said the journalists, there is an advantage in reporting on environmental issues: problems and evidence are out there for you to witness and discover. In order to find out the impacts of the Zijin pollution incident on the environment and who might be the victims, Shang Hua observed the villages along the river that were connected to the mine and the villages that were opposite the mine and interviewed the villagers. In doing so, she was able to obtain intuitive knowledge of events since the occurrence of the pollution incident. Such intuitive knowledge is seen as very useful for her to make judgements about who had been affected by the pollution accident and to what extent the accident has polluted the environment. By doing so, journalists introduce their experience into their reports in order to deepen the understanding of the situation. A result of this is the inclusion of journalists as part of their own stories which reinforces the embedding of journalists' subjectivity in their reports.

As censorship and interference varies from case to case, journalists regard it as crucial to judge governments' attitudes towards the problems and to understand the relationship between enterprises and government. This is a strategy for taking advantage of their different mentalities and conflicting interests. Just as Xiao Feng commented, "the collaboration between local governments and enterprises is not stable. In many ways, local governments have their governmental perspective and standpoint and enterprises have their enterprise perspective and standpoint ... although it is true that many local governments do not want you to reveal their 'skeletons in the cupboard' and set some barriers for reporting because of their concern for political achievement." However, this principle is not always true and really depends on the case. Wang Dong gives an example: "if a locality has an emergent event, for example the discovery of high levels of lead in people's blood, under this circumstance, given the interests [of the local government], you [journalists] would face huge interference from the local government", however, in some other events, "for example, in the recent Bohai Petrol Pollution Incident, the local government would be on the side of journalists because you are expected to constantly expose information, especially the internal information of the enterprise [which is an transnational corporation]."

Making the best use of experts becomes a strategy for investigative journalists. On the one hand, experts act as their think tanks from whom part of journalists' knowledge comes. On the other hand, environmental investigative journalists treat experts and NGOs merely as news sources.

4.7 A case study of "water crisis" reporting (*shui weiji baodao*)

Now I am going to summarise the main points discussed already by analysing a case study of "water crisis" reporting in *Southern Metropolitan Daily*. In 2011, a series of investigative reports on water crises were published in the newspaper. These reports revealed general but severe problems with water quality in China's major rivers, such as the Yangtze River, the Yellow River, the Tuo River and the Huai River, and major lakes, such as the Taihu Lake and the Dianchi Lake. These problems included pollution and its corollary consequences including cancer villages, damaged offshore ecological systems, water shortages, rivers drying up and hollow ground. These ten investigative reports (see Table 4.1) outlined the major environmental problems facing these water sources and the reasons behind these problems. The investigative reporting team at *Southern Metropolitan Daily* spent two months looking into the problems China's rivers and lakes are facing. Their investigation contributed to an overview of the current situation, and the causes and consequences of water crises across China.

According to the journalists, there were two main reasons why the newspaper decided to deploy the whole investigative reporting team on this series. First, water problems became a hot topic from 2005 to 2007; the Songhua River Pollution Incident took place in 2005 and an explosion of chlorella (algae) occurred in the Taihu Lake in 2007. Both these environmental emergencies made water problems a pressing issue (Interview with the former director of the investigative reporting team, 2011). Second, this series was closely associated with the personal experiences of investigative journalists who witnessed serious environmental problems in rivers and lakes with their own eyes. This series originated from the reporting team members' casual chats about water pollution in their hometowns over a dinner. Their chats were extended to a panoramic investigation into China's ten rivers and lakes and therefore the birth of the ten investigative reports in this series (Lu 2009).

Of most interest here is that there is an intimate connection between the origin of investigative reports and the occurrence of environmental incidents or journalists' personal experiences. Almost the whole investigative reporting team participated in this series. Most of the journalists were from small towns and cities outside Guangdong Province. The majority of them were university graduates and several had postgraduate degrees. They have witnessed and have memories of water problems in their daily lives.

In terms of the cost of investigation, on average an investigative report needs about one month to complete and costs around 10,000 RMB. It was said that there was not even a ceiling set for the cost of this investigation. An important reason emerging from my interviews was that the newspaper is a subsidiary and has to give its profits to the parent newspaper and the media group. Therefore it makes no sense for the newspaper to save money in reporting, as the money that has not been spent will be given back to the media group eventually.

Journalists stressed the importance of research and empirical fieldwork in finding out the truth about problems with water quality. Lu Bing, for example, who reported on the water problems in the origin of the three rivers (*sanjiangyuan*) and in the Dianchi, read and researched extensively in order to identify what remained unclear and deserving of his attention in relation to these two topics (Lu 2009). For Li Jing, he and his photo journalist spent nearly 20 days collecting data for their report on the collapse of the offshore ecological system. He did not stop the investigation until he knew what to write about in the report. They walked through all the provinces from the south (Zhejiang) up to the north (Shandong, Tianjin and Inner Mongolia) in order to understand what happened to the ecology along the east coast of China. He used two words "observation" (*guancha*) and "fieldwork research" (*tianyekaocha*)

Table 4.1   Ten investigative reports in the special issue on the water crisis in China
(*zhongguo shuiweiji*) in November 2011

Titles of Reports	Main Themes
Loads of Black Water Flow East, Chain of Chemical Industrial Zones Destroy China's Seas	Seas in danger: Chemical factories have polluted and nearly completely destroyed the offshore ecological system such as that in the Bo Sea Bay.
Diaries of Death in Three Cancer Villages	Residents in danger: Chemical factories have contaminated drinking water, resulting in hundreds of villagers in three villages in Eastern China becoming cancer victims.
Dilemma of Tuo River's Five Types of "Low Quality" Water	Damaged rivers: Chemical factories, energy firms and factories have polluted the river water, destroyed the ecological system in the river and local fishery and led to water drying up and an increase in cancer among local residents.
Fruitless 11-Year Efforts to Clean Tai Lake	The Tai Lake region has been polluted by chemical factories. The pollution led to the explosion of Cyanobacteria. Local governments have spent 11 years trying to deal with it but their efforts are fruitless. This is a typical example of China's lakes.
Dian Lake Dies in Urbanisation	Urbanization and overpopulation in the city of Kunming and the overuse of fertilizers has caused severe pollution in the Dian Lake and the explosion of Cyanobacteria.
Threat of Desertification to the Origin of Three Rivers	Desertification threatens the origin of three rivers: the Yangtze River, the Yellow River and the Lancang River.
Severe Water Shortage in Guanzhong Despite the 800-Li Wei River	Urbanization, chemical fertilizer factories and overuse of underground water caused water problems in the Wei River, including pollution and water drying up. Results of this include polluted vegetables and water shortages.
Overuse of Underground Water, Ground Collapse Worries Huabei	Human activities (sand-digging factories, chemical factories, overuse of underground water) have caused a number of environmental problems (including water resource exhaustion, ground collapse, underground water pollution) for Huabei (Tianjin, Hebei and Shandong).
Collapse of Wastewater Reservoir Upstream from Villages	An incident of collapse of a wastewater reservoir upstream polluted villages and the downstream of the Yellow River.
Pollution Marching West, When Agricultural Irrigation Channels Became Sewage Channels	In the process of rapid industrialization, the west part of China has benefitted from economic globalisation but meanwhile fell victim to pollution transfer from east/south to west.

to describe his work in the process of investigation (Interview, 2011). Before Qu Nian decided what to focus on for her report on the water shortage problem in Shanxi Province, she also did a lot of research and read publicly accessible open data in order to identify what was most important and most interesting to report. Extensive research and fieldwork, however, integrate journalists' personal experiences and observations.

The knowledge involved in the reporting of this series included professional specialist knowledge about relevant environmental problems, knowledge and cognition gained from their personal experiences, and analysis based on all the evidence collected. Experts, ranging from government officials and scientists to NGOs, are one of the main sources of information that helped the journalists to form judgements about the topics. However, journalists share a view that government officials are the news source that is most difficult to access and political interference is without doubt the number one limitation on their reporting.

To take an example, Qu Nian's investigative experience in Weinan City is very representative of this. Qu Nian described in detail how she was prevented from investigating the situation about water and drought in Weinan City, Shanxi Province. She was followed and even "hijacked" by the local propaganda department during the first fruitless 20 days she stayed in the city. Although environmental problems are less politically sensitive than other topics, they still sometimes invite political control. Due to the fact that she lacked social resources, such as people working for corresponding departments in the locale because *Southern Metropolitan Daily* is an out-of-region newspaper, she had to contact local journalists for information. When she attended a dinner with local journalists, she was introduced to government officials who obviously disseminated the information that a journalist from *Southern Metropolitan Daily* had come to the locale to investigate something. This information itself was scary to local officials as *Southern Metropolitan Daily* has a critical image. Local governments feared journalists from *Southern Weekend* and *Southern Metropolitan Daily* and regarded them as potentially troublesome, that is, that they came here to look for, or, more precisely in the minds of the local officials, create, troubles for them. She was therefore monitored by the local government and thus was unable to investigate in the way that she planned to do.

When she found she was blocked from any government information and even kept under surveillance in Weinan City, she employed some alternative tactics. For example, she checked whether there were any papers the officials and experts in the agricultural department and water resources department had published and tried to compare the information in these papers with open data she found on the government websites. She also pretended to leave the city in order to avoid surveillance and control by the local government and asked one of her friends who happened to be there to register a hotel room for her. (We used the same tactic when an investigative journalist and I were investigating the mine disaster in Shanxi Province, 2006.) In order to gather official information, she visited a retired engineer because she felt retired officials or technical experts were usually more willing to talk and had less sense of political sensitivity. In addition, she also was able to find relevant information on governments' websites because officials had to write and publish their reports under the requirement of open governance. They usually place all work-related documents on their governance websites. If she asked officials about the open data on their websites, they sometimes would be willing to tell her relevant information. But still, all the information collected was very fragmented and there was a large amount to digest, to analyse and to organise in a logical report.

The authors of the ten investigative reports in this series and their directors acknowledged the importance of choosing angles and making judgements. Tang Yan, to take an example, thought the two topics he was reporting on were not new at all, as so many people had reported on them before. But he still believed he was able to create unique stories because there were many details and many hidden truths that people did not know and it was his job to reveal them. He found that some important questions remained unanswered in the literature (here the literature means available news reports and academic articles that the public can access; this process is quite similar to what researchers do in reviewing the literature and defining the problem). After he analysed the information he obtained, he felt that the theme of the report on the origin of the three rivers should be around whether and how the state's funds were used for removing pasture and helping to restore grassland and re-establish ecological migrants' lives, while his report on the Dianchi pollution should focus on the problems with cleaning the Dianchi Lake. This was because little research had been done in this field.

An interesting point in relation to the logic behind all these reports and the understanding of environmental problems is that the former and present directors of the team, at the time of the interviews, agreed that it was the "system" (*tizhi*) that caused environmental problems, although different environmental problems had their own special causes. The dichotomy between governments and the environment was deemed as the best approach to take in reports on environmental problems. And because of the embodiment of systemic problems in the state of the environment, reporting on environmental problems can reflect the problems of the system while remaining acceptable to the authorities and politically relatively safe. This principle, of course, has guided environmental reports and has even become a routine framework, as one can identify in the reports.

4.8 Conclusion

The story of Chinese environmental investigative journalists and their work presents a picture in which journalists endeavour to strike a balance between being "professional" journalists and participants in events that inevitably involve their personal judgement, experience and cognition. It is their doubts about experts' and officials' words, their trust in their own judgement and choices of reporting angles that underlie the discursive patterns identified in the coverage of investigative reports in Chapter 3.

As has been shown in the discussions in this chapter, environmental investigative journalism in China is a genre of journalism that is far from being "objective" journalism. Its advocacy nature, together with the prominent involvement of journalists' personal experiences in the reports, results in subjectivity characterising environmental investigative journalism. This offers insights into the features of the discourse of investigative reports. The initiative of journalists to verify and compare information, judgement and the choices of reporting angles influence the meanings and frames of their reports. The frames in their mindsets and journalists' cognition of environmental problems and judgement impact the selection of topics, themes and materials. Individualism embodied in the whole process endows their reports with individual interpretation of environmental problems and specific meanings between the lines of their reports. Individual investigative journalists work as the key agents that select socio-environmentally important events out of perhaps millions of events in the world and define them in certain contexts on the one hand and communicate or translate scientifically abstruse environmental knowledge to the public on the other. Through bringing in subjectivity in reports, journalists are able to manipulate spaces to avoid the potential domination of elite news sources and to negotiate the control of powerful individuals or institutions. Meanwhile they are able to bring alternative discourses on environmental issues and tell their own stories that may be opposed to the dominant discourse promulgated by the party-state. The downside of this, nevertheless, is that the meanings of environmental problems are highly associated with journalists' understanding and experience of these problems at the individual level, as influenced by their backgrounds and the extent of their knowledge. Their stories narrated in investigative reports may also follow routines of their own. Investigative reports have thus become a site where the meanings of environmental problems are made.

This study shows that investigative journalism takes a mixture of an advocacy and informing role in reporting on environmental problems. It is advocacy journalism because it advances an idea of environmentalism that treasures environmental health, sustainability of nature, the harmony between nature and humanity, and environmental and social equality that opposes modernisation, its principles and what it can bring to society. This idea is constructed out of the personal experiences, beliefs and understanding of investigative journalists about the relationship between humanity and nature. This idea springs from the worries of investigative journalists about what they see and the sympathy they feel for victims of environmental destruction. They permeate their investigative reports with what they believe is true and right and that they would like to pass on to the general public and have accepted by them.

They have an informing role because they make their readers aware of the emergence of environmental problems and risks by contextualising these problems and risks. This informing role indeed rings a warning bell for modernisation. It offers a basis for the advocacy role of investigative journalism in reporting on environmental problems. Investigative reports on environmental problems first tell readers about what has happened and then tell them how to understand the events and what to believe.

But what investigative journalism does is far from being objective, given the journalists' initiatives, epistemology and unique news production process, although almost all investigative reporters who have agreed to be interviewed declared that they are objective in reporting. Their claim to being objective can be understood, and interpreted by themselves, as presenting evidence and viewpoints from a variety of news sources in a balanced way so that they can hide their own opinions. Ideally they express the hidden meanings that represent their own opinions through the mouths of their news sources. This is seen as the "professional" practice of investigative journalism.

The next chapter will further look into the interaction between offline investigative journalism and online environmental movements and examine how this interaction helps to forge environmental discourse in Chinese society.

5
Offline Investigative Journalism and Online Environmental Crusades

The current century is a time of online environmental crusades. China's online public space is filled with citizen-initiated environmental discussions and groups, as facilitated by the wide spread and application of the Internet, its associated Web 2.0 tools and other ICT devices. In particular, visual images and texts that can evoke strong emotions among Internet users, such as images of polluted rivers, dead finless porpoises, dense smog, the striking appeals of residents of "cancer villages" for survival and salvation, and reflections on the construction of the Three Gorges Dam are widely circulated online. These first-hand accounts of the damage to the environment from human activity and ordinary people's suffering are detailed, sensational, packed with personal experiences and feelings. They create a space in which the wider public can see what is happening to the environment in distant places as well as the potential risks posed to the whole nation by human activity locally and over greater geographical distances. The capacity to have the public bear witness to environmental destruction and make them aware of environmental problems and risks can resonate among the public and even trigger outrage. The emergence of online environmental crusades on the one hand forges a civil critical voice that opposes governments over the national priority for economic growth and, on the other, demonstrates that it is no longer solely down to investigative journalism to reveal environmental problems and mobilise the public. These two aspects thus pose questions for investigative journalism and its role.

The first question concerns the mobilisation role of investigative journalism. The ultimate aim of investigative journalism is to mobilise the public to influence policies and policymaking (Ettema and Glasser 1998).

139

In other words, investigative journalism exercises its power through revealing information to the public and mobilising the public. Looking at historical precedents, one can see that environmental movements that want to rally the public, who share the same environmental values, to protect the environment, have needed investigative journalism to get their messages out to the public to mobilise them. Such a relationship between investigative journalism and environmental movements, nevertheless, could become shaky if investigative journalism lost its mobilisation function to online environmental crusades. If that is the case, online environmental movements would take over from investigative journalism to be the main agent of social change in relation to environmental problems and governance.

The second question is how investigative journalism positions itself in the opposition between online civil discourse promoting environmental protection and the dominant official discourse prioritising economic growth. Online environmental movements raise the public's awareness of and anxiety over environmental problems and also form a critical civil discourse contesting the dominant Promethean discourse of the environment that favours economic growth over environmental protection. Unlike other genres of news, the adversarial nature of investigative journalism suggests it is likely that investigative journalism will take the side of online users and reject the Promethean discourse of the environment, as it has led to numerous insoluble environmental problems and risks. However, in general, news media in China are expected to play the role of "ideological state apparatus", that is, to help maintain the existing social order. Support for the capitalist mode of production and the priority for economic growth are two of the main ideological pillars that the news media should help to reinforce. The second crucial question for investigative journalism is therefore whether investigative journalism acts to exaggerate or ease the contest between the two discourses and to what extent.

This chapter addresses these two questions, analysing the patterns of the interplay between online environmental crusades and offline investigative journalism. In doing so, this chapter examines the role of offline investigative journalism as an amplifier for enlarging online critical civil voices in relation to environmental problems and issues. The performance of investigative journalism in response to the rise of online environmental crusades offers a perspective to understand the role of investigative journalism in constructing environmental discourses in China. This also provides a prism to investigate the relationship between investigative journalism and environmental movements, a relationship that has changed in recent years.

Environmental movements have presented new features in the new media era and have appeared to have greater public opinion supervision and public mobilisation functions than in the past, while the Internet offers new spheres in which environmental problems and issues are exposed to the public and by the public. Online environmental movements that are discussion-based are able to quickly draw the public's attention and to ignite heated public debates on particular environmental problems and the wrongdoings of individuals or institutions, but often lack sustainability and take place in an unsystematic way. The publication of investigative reports constructs a space where online civil environmental discourse and offline official discourse meet. What Internet users and governments have to say is presented side by side in a particular and allegedly objective way in investigative reports. In addition, investigative journalists' own investigation and research endorse the existence of relevant environmental problems, revealed by Internet users, and can make reference to other environmental problems and even introduce new frames in their reports.

Environmental agendas central to online discussions, nevertheless, do not always enter the agendas of investigative journalism. Investigative journalism retains its control over the news agenda. Investigative journalists decide whether or not to include the topics of online environmental discussions in their reporting. Therefore it would be difficult to say that online environmental crusades determine the agendas of investigative journalism. In contrast, the attention of the online public on environmental problems is maintained by the continuous feeding back of investigative reports into online discussions, though they cannot always attract the attention of Internet users.

Investigative journalism and online environmental movements play their respective roles in two – for most part separate – spaces with the same purpose of confronting political authorities, commercial institutions and individuals in relation to environmental problems. Covering online topics picked up by investigative journalists, investigative reports in mainstream media coverage strengthen online worries over the environment, representing a type of civil discourse in a society that is concerned about the environment.

It is argued in this chapter that offline investigative journalism and online environmental crusades are now increasingly working as intimate partners in advancing official investigations into particular environmental problems and issues. Online environmental crusades put pressure on governments regarding environmental governance and relevant governmental policies. Picking up these online discussions, investigative journalism amplifies this discourse through investigating and verifying the facts revealed by Internet users and escalating their significance by contextualising and generalising them. In this way investigative journalism, in alliance with online environmental crusades, breaks the state's monopoly of discourse on the nature–development relationship and pushes governments to enhance environmental governance.

5.1   Online environmental movements and new environmental spheres

On February 10, 2013, a Weibo post accused many factories in Weifang City, Shangdong Province, of discharging wastewater into 1,000-meter-deep underground water through high-pressure wells and equipment. In this way the discharging of wastewater completely polluted the whole body of underground water. This post was widely circulated in cyberspace, especially after having been forwarded by Deng Fei, a renowned journalist. This event took place during China's lunar new year holiday. Following the exposure of this wrongdoing in Weifang, Internet users who were spending the festival with their families in their hometowns started publishing images and videos of scenes of water pollution in these towns. For example, Wang Wenzhi, a journalist at the Xinhua News Agency's *Economy Reference Daily* (*jingji cankao bao*), was one of these Internet users. Deng Fei and other activists further initiated the "All Citizens Take Pictures/Videos of Pollution Scenes in Hometowns" (*jiaxiang wuran quanmin pai*) campaign on Weibo that called for the public to record and expose the local pollution problems on the Internet. This campaign turned the spark of individual Internet users' concerns on water pollution into the national flame of an anti-pollution campaign. In an interview with a newspaper, Deng Fei expressed his view: "our action does not merely target Weifang, but the whole situation of environmental pollution [in China]. Our aim is to truly represent the problem of pollution and collect relevant evidence in order to raise the awareness of the general public and increase their public opinion supervision capability."[1] Seemingly overnight, everyone started talking about (underground) water pollution. Later on, online discussions even escalated into a "national turmoil" in the words of Zhang Bo, the Director of the Environmental Protection Department in Shandong Province.[2]

Local governments in Weifang and Shandong Province, nevertheless, denied the existence of underground water that had been polluted by wastewater from factory discharges. They implied that Internet users, including Deng Fei, were disseminating untrue rumours and requested

Deng Fei to apologise on Sina Weibo to the local enterprises involved.[3] A fierce debate followed, in which both local governments and Deng Fei expressed their views and raised questions for each other on Weibo. However, the response of the local authorities further triggered the anger of Internet users who suspected collaboration between local authorities and enterprises. As a result, although local governments insisted there was no such thing as underground wastewater discharging, most internet users still believed it was true.[4]

Leaving aside the disagreements between local governments and Internet users, one thing was certain: this event indeed drew national attention to the issue of underground water pollution. The Ministry of Environmental Protection of the PRC, for example, carried out an inspection into the discharge of wastewater in six provinces in North China from the second half of February to March, 2013.[5] The six provinces were at the centre of the accusations by Internet users. The inspection revealed that 55 enterprises in these regions illegally and inappropriately discharged wastewater. Among the 55 enterprises were several enterprises in Weifang which refuted the claim of the local governments of Weifang and Shandong Province that no local enterprises had illegally discharged wastewater and polluted underground water.

Another important consequence of this event is that some local governments adopted the "All Citizens Take Pictures/Videos of Pollution Scenes in Hometowns" campaign to encourage the general public to participate in environmental protection and exercise public opinion supervision. For example, later in 2013 the Environmental Protection Bureau in Shandong Province invited citizens to join its "Taking Pictures/Videos of Smog Pollution Close to You" (*fencheng wuran shuishou pai*) campaign. This official campaign attempted to boost the participation of citizens in monitoring air pollution from every corner of society. This campaign was said to be successful and the bureau even gave awards to four citizens for their contribution to revealing pollution.[6]

This case of the campaign over underground water pollution in 2013 is of course only one among many other environmental movements appearing online in recent years. Along with the fast-growing popularity of Weibo since 2009, the emergence of online environmental movements in China has attracted global interest. The importance of the Internet to environmental movements had already been recognised at the outset of the 21st century. Some scholars, such as Yang Guobing, see the Internet complementing and making up for the flaws of NGOs and news media in mobilising the public (Yang 2003a). The proliferation of the Internet facilitates public participation in environmental debates.

ENGOs have set up websites and Weibo pages disseminating relevant information, and initiating and participating in discussions on environmental problems. Investigative journalists and news organisations not only turn to the Internet for enormous sources of news but also broaden the channels in which their reports can be disseminated.

Online environmental movements are largely electronic-based and discussion-centred, typically taking the gentle form of public debates and discussions, with the exception that on some occasions physical protests on the streets are involved. Online campaigns of this type can involve Internet users revealing cases where individuals and institutions have harmed the environment and environmental destruction that they have personally witnessed. Often they express an expectation that government will solve the problems and articulate a longing for a clean and safe environment. They are characteristic of unsystematically spontaneous public participation, which is different from the elite-organised movements in the 1990s. Environmental movements of this kind involve both elites and the general public, who are relatively equal in the face of universal environmental problems, but unequal in terms of their positions in the network, the resources they access and the scale of their influence. Elites often play –or attempt to play – a significant role in escalating specific events into general national issues, as manifested in the case of the Shandong Weifang underground water pollution in 2013. In that specific case, Deng Fei's re-tweeting of the Weibo post and advocacy over the issue was a turning point and took the event to a national level.

Among others, prominent environmental movements have included the campaigns against PM2.5, underground water pollution, cancer villages and the construction of rubbish-burning projects, of power stations and of dams and hydroelectric projects, the campaigns promoting finless porpoise protection, and the campaigns prompted by the frequent occurrence of environmental incidents. In the early 21st century many of these agendas only captured the attention of the public after an extensive investigation by journalists and the publication of their reports. By contrast, nowadays it is not unusual to see these agendas raised by the public on the Internet preceding the publication of any investigative reports in the mainstream news media.

These agendas are of two types. The first type is environmental issues which are not restricted to particular localities. They have a national scope and are geographically dispersed. Being national in nature means there are no specific targets for criticism, although in some cases local governments, organisations and individuals might be criticised.

Prominent topics in this agenda include underground water pollution, cancer villages, the extinction crisis of finless porpoises, PM2.5 and smog. Though national in scope, specific local events in relation to these issues, such as an occurrence of pollution, later resonate nationwide.

The second type of topic has clearer targets for criticism than the first one, usually the policies of governments at all levels in relation to construction projects such as hydroelectric and nuclear plants, dams, incinerators and the Nan Shui Bei Tiao project (the project to divert water from the south to the north). Online environmental movements on these topics often occur as a result of government decisions on constructing certain projects that worry the public, or when some natural disasters, such as earthquakes, droughts and extreme heat, hit specific regions of China where these projects have been built or are planned. These movements are much more critical than the first type, since they directly confront government policies including those of the central government, and question their correctness.

Apart from these two types, there has been a proliferation of street protests during this period, including those against PX projects in Xiamen (in 2007) and Ningbo (in 2012), the Shanghai Underground project (in 2009), incinerators in Guangdong (in 2009, 2013 and 2014) and in Yuyao (in 2014), a coal-fired power station in Haimen (in 2011), the construction of wastewater pipelines for a chemical factory in Qidong (in 2012) and a copper plant in Shifang (in 2012). These environmental protests share several prominent features. They are usually local and occur in specific regions, but fail to spread to other places. They are usually triggered by local governments' decisions to launch industrial projects. They are often mobilised and organised on the Internet, that is, participants disseminate information about the projects and organise their actions on the streets by using new media technologies and Web 2.0 tools, such as mobile SMS, Weibo, and QQ. There is no obvious participation or organisation by NGOs. They commonly move to offline street protests after an online mobilisation. In addition, the news media generally do not report in advance of environmental protests of this kind, due to media control, let alone publishing investigative reports. Any in-depth reports published are likely to be post-event reviews and analyses of these protests.

In these new environmental movements, we can see elites and NGOs trying to take control and act as opinion leaders guiding the development of these movements. In cyberspace, members of elites (who often have a huge group of followers on their social media accounts) are frequently journalists such as Luo Changping, Bai Yansong and Deng

Fei, business people such as Pan Shiyi, Xue Manzi, celebrities such as Yao Chen and politicians such as Cai Qi and Wu Hao. They have attracted a large group of supporters/followers, are very influential and even more persuasive than NGOs.[7] A single tweet produced or forwarded by them might be able to trigger a national wave of public sentiment, as exemplified in the case of the Weifang underground water pollution and in the case of PM2.5. Deng Fei's re-tweeting of the post of an Internet user was the crucial moment in making this a national cause. Likewise, in the case of PM2.5, in 2011, after the American embassy published figures on PM2.5 that were different from the ones published by the Chinese government, Pan Shiyi, a well-known real estate developer and the CEO of SOHO, started monitoring PM2.5 and publishing the results he got on his Weibo page (he had 7.4 million followers at that time, which rose to 16.7 million by March 2014). He also initiated an online-based vote asking whether or not the Chinese government should adopt the international standard index of PM2.5 and publish the results.[8] The result showed a landslide supporting the idea (32,000 micro-bloggers agreed, while only about 250 disagreed, along with around 120 who weren't sure).[9] Not much later the Chinese government agreed to publicise the PM2.5 index information from January 2012.

Elites on the Internet, especially journalists who have access to media resources, try to collaborate with investigative journalists, environmental activists and NGOs to mobilise the public and to investigate the status quo of environmental deterioration. In the case of the Weifang Underground water pollution, for example, Deng Fei initiated a campaign called "Independent Investigation into Chinese Water Crisis",[10] attempting to gain the support of investigative journalists, environmental activists and NGOs nationwide.

However, there is not enough evidence to establish the prominent role played by elites in the discussions on, and environmental protests against, the construction of certain anthropogenic projects. The role of members of elites and ENGOs was even missing in some environmental movements such as in the Yuyao protest against the incinerator project in 2014. The degree of passion of elites to be involved in discussion is associated with the nature of these types of environmental problems. Environmental problems caused by hydroelectric projects or incinerators, for example, are geographically constrained and thus have little impact on elites who are not based in those areas, while air and water pollution are universal and no one (including elites) can escape being affected by them.

5.2   The political implications of online environmental movements

Apart from the direct confrontation between Internet-facilitated environmental protests and political authorities, the challenges presented to the government by the emergence of online environmental movements come not only from the collective resonance Internet users achieve when discussing environmental issues, but also from the construction of critical environmental discourses that oppose the dominant official Promethean discourse and push governments to make policy changes accordingly.

The Internet functions as a platform to share and reveal information about environmental problems and issues of which the general public are otherwise unaware. Online discussions on environmental issues and information communicated on the Internet often involve pictures and even videos as well as accounts of the personal experiences and feelings of Internet users. Symbolic information about current environmental situations and about environmental victims make individual Internet users aware of what is happening to the environment even at a great distance and enables them to exchange personal experiences and feelings. The sharing of information about the environment turns the Internet into a space where Internet users can recall their own childhood memories of the clean environment and construct collective memories, bearing witness to influential environmental incidents and triggering collective resonance about problems happening both where they live and in distant places, and a place to express worries for the survival of humanity.

Collective resonance is constructed among Internet users who come together online to share similar experiences and concerns over the environment and the future of their homes. Ordinary people might know what is happening in their own local surroundings. When they pass by a river running through the cities in which they live, they might be able to see and judge the quality of the water by observing the river. However, they are not able to judge whether this is an individual case or a general problem facing China. Once such images and videos about water pollution – whether in cities or remote areas such as villages – are uploaded onto the Internet, however, they constitute a more complete picture of the situation in China as a whole. Internet users can compare their immediate reality to a distant reality that may or may not impact on their daily lives.

Collective resonance in part also originates from Internet users' nostalgia for their hometowns and the past. To take an example, the "All Citizens Take Pictures/Videos of Pollution Scenes in Hometowns" campaigns show Internet users' nostalgia for the past and worries for the present and future of their own hometowns. In this specific case, it is easy to identify this kind of nostalgia in the comments of Internet users. For example, an Internet user posted his/her comment on water pollution: "I still can remember when I was a kid I played with mud and sand on the riverside and my mum washed clothes in the river. However, now the river has lost its natural mud and sand and instead is full of rubbish. The river water is smelly and the fish have died out. Now we and the river can no longer go back to the past!" This comment was made in a response to the post of another Internet user: "The river in my hometown was clear when I was young. Now I have grown up and the river is gone. ... "

Encountering online stories about the miserable lives of environmental victims prompts the humane sympathy of Internet users for these victims on the one hand, and pushes them to reflect on the environment they live in and on the possibility of being victimised themselves by the damaged environment on the other. A prominent example is online appeals by cancer villages' residents. Their appeals directly bring their miserable experiences to the public without depending on the mediation of traditional media. They are stuck in remote villages that are suffering from the side effects of unsustainable industrialisation because of their lack of resources to change their fate. Poverty and disease are the twin fates of these peasants who have been victimised in the hope of getting a better life but were not even aware of the potential risks posed by the operation of factories on their doorstep. Through appealing online, those who share a similar doom huddle for warmth and strength in order to face an unpredictable future that is out of their control. On the other hand, these appeals also trigger other Internet users' worries over the survival of humanity in a damaged environment and win their humane sympathy, which may or may not lead to changes in their fate.

As Doyle and McEachern have commented, "environmentalism in all its forms, was born in environmental movements" (Doyle and McEachern 2007: 84). Collective resonance about the environment helps forge the prevalent environmental discourses that portray an image of a damaged environment and of suffering human beings and animals. On the one hand, online discussions represent the environment as sick, damaged and vulnerable to human activity. Pictures and videos about polluted air, lakes and rivers present a miserable image of the environment. Textual descriptions of environmental crisis, such as the drying out of the lakes downstream of the Yangtze River that has the Three Gorges Dam in the upper stream, highlights the link between the damaged environment and human activity. On the other hand, the suffering of humans and animals, such as the gloomy lives of inhabitants of cancer villages as well as the death of finless porpoises, depict another terrifying picture of the environment taking revenge. This picture implies that humanity as a whole, which destroys the environment, will bear the consequences of the damage it has done, though often it is the underprivileged people and social groups who suffer from the side effects of modernisation.

These environmental discourses present a critical opposition to the national priority for economic development and express the desire for a clean and safe environment. A poll published by Tencent News in February 2013, to take an example, invited Internet users to choose between economic growth and environmental protection. At the time of writing (March 4, 2014), 30,905 Internet users voted for environmental protection with only 531 users for economic growth.[11] This civil discourse presents a challenge to the dominant national discourse of economic development.

The ultimate dual effects of revealing information are to form public sentiment for a clean environment as well as to provide information to governments who are expected to do something to stop the ongoing environmental damage from happening. With the facilitation of the Internet, information about environmental incidents and protests can be rapidly disseminated beyond the limits of time and space and beyond the control of the political authorities. The importance of information dissemination of this kind lies in increasing the visibility of these events, which in turn invites public surveillance over how the events are handled. Information of this kind breaks news to the general public in the first place and continues to inform them about events in their own locality and elsewhere, even when traditional news media have lost interest or are unable to report on these events. This is exemplified in the cases of cadmium pollution in Liuzhou, phenol pollution in Zhenjiang and aniline leaks in Handan (in 2012), and in environmental protests such as the protests against the coal-fired power plant in Haimen in 2011 where there were very few reports published in traditional mainstream media at all. On these occasions, relevant information being circulated on the Internet greatly increased the public visibility of the events. Likewise, in the phenol pollution incident in Zhenjiang in 2012, a simple search on Sina Weibo came up with 57,680 posts mentioning this incident from

October 14, 2012–2013. Although most of these posts are re-tweets of a post by an environmental activist, and do not reveal much about the incident or its consequences, the number of posts contrasts sharply with the fact that during the same period only nine news articles mentioned this incident in all the Chinese newspapers that are included in the Wisers Database.

The effects of online environmental crusades are thus reflected in public opinion and public mobilisation. The online campaign over PM2.5 in 2011, discussed above, is an excellent example of this. In this case, first the American embassy in Beijing and then economic and cultural elites published their own PM2.5 figures on Weibo, which were widely forwarded. This forged a strong reaction in public opinion that demanded that governments increase transparency about air quality. Under the pressure of public opinion, the central government was pushed to respond.

Although the overall online attention to environmental problems continues, online environmental discussions on particular topics are often transient. Discussions of this kind are generally triggered by particular posts, events or news reports. The attention of online users both aggregates and disappears quickly. Take the case of underground water pollution for example. According to Google trends and Sina's Micro-Index (*weizhishu*), this agenda suddenly reached a peak in February and March 2013 (when Deng Fei and other Internet users raised the question) and dropped sharply afterwards. There was only another minor peak later on in December 2013. During the rest of the time, little attention was paid to this issue. Similarly, in 2014, according to the Sina's Micro-Index, the online attention to the agenda of "rubbish burning" on Sina Weibo started to rise from May 9, reached a sharp peak on May 11, the next day of the Yuyao anti-incinerator protest, and then declined in an abrupt way. The attention disappeared on May 15. Therefore, the online attention triggered by the Yuyao protest in 2014 only lasted for about one week.

It is obvious that online discussions on particular environmental topics are driven by actual events in society. This point is especially clear in examples such as the construction of the Three Gorges Dam. For this subject the trend is complicated and closely coincident with the emergence of environmental and socio-political events. From 2004, peaks appeared in May 2006 (at the time when the project was finished), May 2008 (the Wenchuan earthquake occurred in that month), May 2011 (the Chinese government officially acknowledged in a statement that the project had caused certain damage to the local environment) and

April 2013 (the Lushan earthquake hit Yan'an City). In addition, there was an increase in March 2014. On February 27, 2014, an investigative report in *Times Weekly* (*shidai zhoubao*) revealed corruption in the bidding process for the Three Gorges Dam construction project and that some "retired cadres" had interfered in the open bidding ("Liuyuan: Exclusive revealing the profit chain behind the Three-Gorges Group"), implying that some individual officials in important positions had benefitted from the construction and suggesting that it would be a good thing to find out who they were. Shortly afterwards, Li Peng was named in some online posts, suggesting that he should be held responsible and may be the central government's next anti-corruption target after Zhou Yongkang. For this topic of the Three Gorges Dam, both events and disasters and investigative reports have triggered peaks in the discussions and have helped to keep online attention on this topic over a long period of time.

Of course, opinions in discussions on the Internet are never one-sided. Social actors, collectively and individually, with respective interests, employ new media technologies to disseminate messages to the general public online. In the case of the underground water pollution in Weifang, for example, an environmental protection bureau in a city in Shandong Province posted 14 messages on its Sina Weibo account in March and April 2013, expressing its determination to deal with local pollution and the Environmental Protection Department of Shandong Province and the Environmental Protection Bureau in Weifang City also used their official pages on Sina Weibo to refute the accusation made by Deng Fei and other Internet users. At the same time, local water-cleaning technology companies kept promoting their water-cleaning products on their online pages. When rubbish burning became a major topic, the Weibo page of China's Electricity News Net even published a post that praised the contribution rubbish burning made to generating electricity. Multiple social actors are involved in discussions on topics such as the construction of the Three Gorges Dam. In 2011, to take an example, the Three Gorges Dam Group published several statements on its official Sina Weibo account denying any relationship between the Three Gorges Dam and drought and urging the news media to positively report on the dam. Therefore, the Internet itself is a neutral medium that provides a platform for anyone who has the willingness to use it to advance their own interests. They speak in different voices that forge distinctive discourses of the environment. But what is more important is which discourse will be picked up by the general public as well as the news media – especially investigative journalists – and allowed to reach the mainstream media discourse that connects official discourse and civil discourse.

The political implications of online environmental crusades are also limited by the dichotomy of villages versus cities and the inequality between urban and rural areas. There is an obvious mismatch between the geographical base of Internet users and the locales where many of these specific environmental problems happen. Generally, Internet users based in rural areas only made up 27.6% of the total population of Internet users in December 2013.[12] That is to say, most Internet users are based in urban areas. However, many environmental problems, including pollution, cancer villages and those caused by the construction of anthropogenic projects like the Three Gorges Dam, take place in rural areas (Wang 2010; Chin and Spegele 2013). It is questionable to what extent and for how long these rurally based environmental problems can attract the attention of urban-based Internet users.

5.3   Investigative journalism's attention to online environmental movements

Investigative journalism has paid attention to the environmental topics discussed online. There is an interesting match between the peak of online discussions and that of investigative reports on particular environmental problems. Two examples that will be discussed in detail here to explain this match include: the online campaigns on the extinction crisis of finless porpoises and on cancer villages. The former represents the fight for nature, while the latter the fight for well-being.

Porpoises live in the freshwater of the Yangtze River and the lakes along the river, such as the Dongting Lake and the Panyang Lake. The size of the population has been decreasing from around 2,700 in 1991 to half that number by 2006 with an annual death rate of 5%. Especially from 2008, every year has seen the death of around 20 porpoises.[13] It is said only around 1,000 porpoises are left in these rivers and lakes at the present time. Causes for the porpoises' death include being hurt by sand-digging boats and instruments and by the methods used to catch fish, such as electro-fishing or blast fishing, poisoning by polluted water, starvation due to food shortages as a result of water pollution, and low water levels caused by the construction of dams and so on.[14]

The appearance of intense investigative reports on the extinction crisis of porpoises immediately followed the surge of online attention to this issue in 2011, although news media including *Southern Metropolitan Daily*, *Southern Weekend* and *Xinmin Weekly* magazine had already published investigative reports on this subject before 2011. Some non-government organisations and associations focusing on porpoise protection have been established since 2008, such as Qiqi's Heaven.[15] Nevertheless, online users did not start paying extensive attention to this issue until 2011. According to Google trends, online attention grew from 2011 and reached a peak in 2012. In 2011, a picture entitled "crying porpoise" published by the Xinhua news agency was widely circulated online. The year 2011 was when the role of Weibo became prominent as an important medium for public opinion in Chinese society. A single search for the keywords "crying porpoise" on Sina Weibo produced 9,888 hits in 2011. According to the in-house record of Sina Weibo, only 908 out of the 9,888 posts contained dissimilar content, whereas the rest were of that picture. It was reported by scientists and citizens that more than 49 porpoises had died within the single year of 2012 (22 died within the first three months in 2012, which is almost equivalent to the number [21] of porpoises that died in the whole of the year 2011). It was also predicted that the porpoises will be extinct within 15 years. In the first quarter of 2012 there was a peak of online discussions on the crisis. Internet users expressed their sadness on Internet forums and Weibo. They also uploaded onto the Internet the pictures of porpoises they had taken and information they found about the situation of porpoises and the causes of their deaths.[16] They also questioned the reasons given by officials for the death of finless porpoises. For example, an Internet user posted a message on Sina Weibo comparing the different reasons for the death of Dongting Lake finless porpoises given in two reports published by *Xinhua.net* and *Yangtze Daily* at about the same time and suggesting that either the officials or experts or both had lied to the public.

Correspondingly, investigative journalists paid much more attention to this issue in 2011 and 2012 than in other years. There are only three investigative reports in 2010 and two in 2009 on the topic of the extinction of finless porpoises. Most news outlets practising investigative journalism (such as *Southern Metropolitan Daily*, *Southern Weekend*, *Xiaoxiang Morning*, *Beijing News* and *Xinmin* magazine) published investigative reports on the crisis in 2011 (17 investigative reports in total) and 2012 (23 investigative reports in total). However, after 2012 the number of investigative reports dropped quickly in 2013 (8) and 2014 (none so far). These investigative reports have usually treated online discussions as a background element for explaining the extinction crisis porpoises are facing. Famous Internet users and blogs/websites/Weibo accounts, such as Haiwenbo and his blog "Qiqi's Heaven" are repeatedly mentioned in investigative reports, as in *Southern Metropolitan Daily*'s investigative report "Counting down for porpoises: an environmental catastrophe behind the collective death of 'National Treasure in Water'" (May 23, 2012) and "Bidding farewell in smiles" at CCTV's *News Probe* (July 6, 2013), as symbols of citizens' efforts to save porpoises.

The case of cancer villages also demonstrates the relationship between investigative journalism agendas and online discussions. The news media had already started reporting on the problem since the end of the 1990s, long before the rise of online attention to it, and important investigative reports on cancer villages began to appear from 2004. Cyberspace has been filled with discussions about cancer villages over the past ten years.[17] This topic is often associated with that of pollution, that is, cancer villages are thought of as a side effect of pollution. 2011 and 2013 have seen the most heated online and media discussions yet on the topic of cancer villages have been seen in 2011 and 2013. In 2011, the media exposure of cancer victims in Tianchang Village, Anhui Province in May, the chromium pollution incident in Qujing, Yunnan Province in August and the relationship between soil pollution and multiple cancer villages across China reported in October have all set off online discussions about cancer villages. Unofficial cancer village lists and maps, immediate information about pollution incidents after their occurrence and information about hidden pollution situations have been appearing on the Internet since then. In 2013 the central government for the first time admitted the existence of cancer villages. Meanwhile a map produced by NGO members became popular online, suggesting that China had more than 200 cancer villages in its territory.[18] According to Google Trends and Baidu Trends, these triggered peaks of online attention to cancer villages in 2011 and especially in 2013.

Correspondingly, the number of investigative reports in these two years are also the highest with 42 in 2011 and 43 in 2013, as compared with 13 in 2010 and 8 in 2012. Like the responses to online discussions on the extinction crisis of porpoises, some investigative reports used online discussions on cancer villages as background information or news hooks. The investigative report "Sequential exposure of cancer villages, water pollution is culprit" in *Economy Reference Daily* (August 12, 2013) is a typical example of this. This report uses the online discussions on cancer villages as a reason for its journalists making an investigation into the topic. However, other investigative reports such as "The death of Huaihe water" in *Beijing News* (June 28, 2013) and "Cancer villages behind nasal bleeding" in *Yangchen Evening* (September 16, 2013), do not mention online discussions at all, but merely rely on the investigation of their own journalists.

However, this does not necessarily mean environmental investigative journalism is driven by online environmental movements. Investigative journalism may not devote any coverage to online discussions on environmental problems or may include diverse topics in its reports. Compared to online environmental discussions, investigative journalists enjoy convenient access to multiple resources and can aggregate information obtained from multiple sources in their reports. The orthodox news-making process adds more credibility to investigative reports that are often thought of as legitimate and reliable. However, investigative reports are limited by editorial policies and routines, such as bans, report timings, news values and so on. Therefore, some topics can be discussed online but cannot be reported as investigative reports in mainstream news media. Investigative journalists need to wait for the right time to report on these topics.

A prominent example of this is the reporting on the connection between drought and the Three Gorges Dam. Suspicions on the connection between the dam and environmental problems – especially extreme weather and natural disasters – in specific areas in China began as early as 2006 when extreme weather and natural disasters such as earthquakes and landslides occurred frequently. However, it was not until 2010 and 2011 that investigative journalists started reporting on this problem. The reporting started because, in the two years 2010 and 2011, the central government expressed different attitudes towards the Three Gorges Dam. For example, in May 2011, the State Council of China issued the Post Construction Work Plans for the Three Gorges Dam (*sanxia houxu gongzuo guihua*). This work plan pointed out for the first time that the Three Gorges Project had influenced the transportation, irrigation and water supply in the middle and downstream of the Yangtze River to a certain extent, and requested local governments to deal properly with these negative consequences of the water storage by the dam. This was thus the right time for investigative journalism to report on the issue of the Three Gorges Dam.

In addition, investigative reports on environmental problems do not limit their topics to specific environmental problems discussed online. Instead, we can find diversity and plurality in terms of the topics of environmental investigative reports. In other words, investigative journalists protect their agendas against online discussions on environmental problems. To take 2013, for example, online discussions as to environmental problems in that year were dominated by the agendas on underground water pollution and cancer villages. However, important news outlets that favour investigative reports such as *Southern Metropolitan Daily*,

Beijing Youth, Beijing News and *Southern Weekend* published investigative reports on diverse topics rather than being limited to these two. *Southern Metropolitan Daily*, for example, published 15 investigative reports on environmental problems in 2013 on seven topics, including the shrinking of wetland, the damage caused by the construction of hydro projects to two rivers, the Nu River and the Dadu River, the quality of drinking water, the pollution produced by factories, cancer villages, silicosis and PM2.5, while *Beijing Youth* covered eight investigative reports on pollution (air, soil, cadmium, river, underground water), the problems caused by waste battery pools, ground collapse and the food security panic caused by concern over possible arsenic-polluted forage for pigs.

The remainder of the chapter will deal with three main aspects of how investigative journalism responds to online discussion on environmental problems and issues. The three aspects suggest investigative reporting brings in the voices of the people to their reports on the one hand but retains its occupational independence on the other.

5.3.1 People's voices visited and official voices invited

By attributing space to the topics of online environmental movements, investigative reports offer a place where the voices of online users and governments meet. Investigative journalists allow Internet users, governments and experts to speak in their reports and carefully construct the views of every side and the facts obtained from multiple resources in a way that they regard as being objective and balanced. A normal way to organise investigative reports on online discussion topics is to present various viewpoints from diverse news sources including governments, experts and ordinary people, such as victims, after introducing the main arguments of Internet users. In doing so, investigative journalists verify the words of Internet users, introduce the response of governments and experts to the questions raised by Internet users and present different viewpoints for readers to make their own judgements.

This is exemplified well by topics such as the connection between drought and the construction of the Three Gorges Dam. As discussed above, online discussions began questioning whether there is a causal relationship between the Three Gorges Dam and droughts from 2006 when Sichuan Province and the city of Chongqing suffered extreme heat and droughts over that summer. However, it was not until 2011 that investigative journalists responded to the debate for the first time. In 2011, the regions in the middle and downstream of the Yangtze River suffered a severe drought, which led to debates, not only among Internet users but also among scholars, on whether the dam was the main cause. The prevailing view among Internet users is that the construction of Three Gorges Dam should be held responsible for the severe drought. Important investigative media outlets such as *Southern Metropolitan Daily* and *Southern Weekend* assigned their investigative journalists to report on this debate. In *Southern Metropolitan Daily's* investigative report entitled "Lakes and rivers in danger: drought and debates in the 'post–Three Gorges Dam' era" (June 1, 2011), Tian Fei, the investigative journalist, cited the arguments of an Internet user "A-Guang" as the most representative of the prevailing view among Internet users. "A-Guang" in his blog article argued that the unscientific way in which the Three Gorges Dam stored water had led to the reduction in water in the middle and downstream of the Yangtze River and therefore the reduction in rainwater. This article was widely circulated on the Internet and its view was regarded as representative of the views of Internet users. Tian Fei also cited the supporting view of an expert whose opinion was that the construction of the Three Gorges Dam had broken the original balance in the natural environment in the region of the Yangtze River and therefore was the major cause of the drought. An opposing view from another expert and a government official's response to "A-Guang's" argument were introduced immediately afterwards in order to show the objectivity of the report.

In the investigative report "Three Gorges Dam tested in droughts" (June 2, 2011) at *Southern Weekend*, Bao Xiaodong, the investigative journalist, did not name any Internet users. Instead, he used "private individuals who care about the Three Gorges Dam" and related their views about how the Three Gorges Dam had caused the droughts. Then he presented the response of relevant government officials that attempted to refute the arguments of these "private individuals" as well as various experts' views in his report. The journalist finally used the words of experts who regarded the drought as having been caused by the dam to refute another expert's opinion that disagreed with the arguments of the "private individuals".

In this seemingly objective way, both investigative reports in the two newspapers presented the voices from the people as well as invited voices from the government. Investigative reports thus become a site where civil discourse and official discourse about particular environmental problems encounter each other. This provides a legitimate route for civil discourse to connect with official discourse and enter the wider domain of public affairs.

5.3.2　Exposing environmental problems in their own investigation

Presenting different views by Internet users, experts and officials constructs a balanced and neutral report. However, the introduction of investigative journalists' own research and investigation tends to disrupt this balance. Some scholars such as Zeng Xufan (Zeng 2009) have argued that news media play a mediating role in the confrontation between online civil discourse and orthodox official discourse about the environment. However, this study has found that, reflecting its adversarial nature, investigative journalism amplifies critical voices in online civil discourse, although investigative reporters believe their aim is still to help resolve the problems, facilitate environmental governance and ease the conflict between the public and governments, rather than confronting authorities (based on interviews).[19] As discussed in the previous chapter, a routine practice of environmental investigative journalists is to conduct research in fieldwork and to include their personal witnessed scenes in their reports. Investigative journalists' own investigation of online civil discourse, but also the references investigative reports make to other environmental problems, imply the side investigative journalism has chosen.

I would like to return to the Shandong Weifang underground water pollution case. This case embodies a clear opposition between Internet users and local governments. Some 22 investigative reports on underground water pollution were carried by newspapers across China in 2013. Rather than clearly acting as an arbiter, judging and deciding who is right and who is wrong, investigative journalists on the whole simply reported on the general situation of underground water quality and the consequences of water pollution. Through journalists' own investigations, the content echoes and enlarges the concerns reflected in online discussions on the issue. "The decline of underground water in China" in *Beijing News* (February 24, 2013) and "Who is to blame for underground water pollution" in *Xin'an Daily* (February 22, 2013) are two prominent examples of this. The former investigated the general situation of underground water quality and the latter also investigated the situation in another city in Shandong Province and confirmed the existence of the phenomenon described in online discussions about pollution in the Weifang case. Apart from generalising the local environmental problem, the investigative reports also made reference to other environmental problems, particularly the topic of "cancer villages". Both the generalisations and the links between underground water pollution and "cancer villages" confirm the concerns reflected in the online discourse about the Shandong Weifang underground water pollution and more generally the quality of underground water in China.

Going back to the two investigative reports on the connection between droughts and the construction of the giant Three Gorges Dam that were discussed above, both journalists opened their reports with their own observations and the testimony of witnesses in the drought-hit regions such as Hubei Province. Journalists described in detail what they saw in the lakes and rivers, for example, the drying out of the lakes and the extremely low levels of water in rivers. Their observations confirmed the problems from which the particular region was suffering, just as had been described in online discussions. The *Southern Metropolitan Daily* journalist even cited the arguments in a research paper to refute the official argument which disputed the connection and to support the online argument that suggests the existence of such a connection. In this way journalists amplify the critical voices online about the negative consequences of the Three Gorges Dam.

5.3.3 Narrative frameworks and diverse frames

Investigative reports of this kind follow a continuous template in narrative frameworks. Following this template, new frames and discourses emerge, which demonstrates that investigative reports are far from being driven merely by online discussions. In the case of the extinction crisis of finless porpoises, three pre-eminent themes (frames) emerged in the nearly 60 investigative reports from 2008 to the time of this writing (March 2014): finless porpoises are dying and facing extinction; human activities and environmental problems are the main causes for the death of finless porpoises; and only human beings can and should do something to save this species. Most investigative reports are structured following a pattern: the death of finless porpoises is introduced at the beginning of the report, followed by a review of the general situation of this species, and they conclude with the analysis of reasons to account for the problematic situation of porpoises given by experts, fishermen, volunteers or even using journalists' own investigations or scenes they have personally witnessed. The construction of dams, illegal fishing, water pollution, sand digging, ships and the callousness of human beings, especially officials, should be held responsible for the situation of finless porpoises. These reports portray an image of the conflict between finless porpoises and human beings.

A similar picture can be found in the case of cancer villages. There is a common template for investigative reporting on this topic. The narrative structure of "Wenyuan Shangba Village: Salvation and Hope" in

Southern Metropolitan Daily (November 18, 2005) by Chuanmin Yang offers a typical example. It opens the narration with a description of how many villagers have died of cancer in the "cancer village" and turns to introduce the cause: the polluted river water in the Hengshi River. The journalist's own experience of the water quality and the horrible lives of villagers as well as villagers' accounts of how poisonous the river water is and how their lives have been affected then follow. Afterwards, the report starts to reveal what has polluted the river water and who should be held responsible for the suffering of villagers through the analysis of experts, the comments of villagers and the scenes the journalist has witnessed with her own eyes. The report then concludes with an analysis based on the comments of experts about the possibilities for cleaning up the polluted river water and the surrounding land in order to save the "cancer village". This report is also accompanied by a commentary written by the journalist which suggests that the management model of allowing pollution to occur before enforcing regulation does not work well. This is an example of an investigative journalist's advocacy role in promoting changes in environmental governance.

This investigative report, published a few years ago, offers an example of a reporting template for investigative reports on this topic. Its narrative framework continues to be adopted by environmental investigative reports in recent years. The investigative report "The death of Huaihe water" in *Beijing News* (June 28, 2013), for instance, by and large follows this typical template, portraying the tragic image of a cancer village and the connection between the cancer village and water pollution, although the journalist has not included any commentary with this report. In this way (following the orthodox narrative structure), investigative reports eliminate the influence of online discussions but instead continue with their traditional narrative frameworks. In the process, diverse frames are added.

There is therefore a general pattern explaining how offline investigative journalism responds to online environmental movements and discussions. Investigative journalists carry out investigations into individual cases discussed by Internet users or the questions raised by Internet users that are deemed as having newsworthiness. In reporting individual cases, investigative journalism generalises individual cases into universal problems, making reference to other environmental problems. Investigative reports often interpret in-depth and multiple meanings of individual cases, upgrading the meanings of topics discussed online by using diverse frames. In addition, investigative journalists also integrate the environmental problems addressed by online discussions into their long-term agenda-setting process. However, online agendas do not dominate the coverage of investigative reports. Instead, more diverse topics of investigative reports on environmental problems can be identified in mainstream media coverage.

5.4 Engaging Internet users

Investigative journalism lacks effectiveness if it fails to engage readers – Internet users here. The feeding back of investigative reports to online movements is crucial. This is because the circulating of investigative reports by Internet users as well as the triggering of online discussions on the topics covered by investigative reports enlarges the effects of these reports on the one hand and demonstrates environmental citizenship on the other.

One can identify the evidence of whether and how Internet users are engaged to discuss the environmental problems raised by investigative reports. The publication of an investigative report on an environmental problem often triggers intense online attention to this problem. The recent reporting on arsenic pollution and its association with cancer villages, for example, demonstrates this effect very well. On April 13, 2014, *Law Evening News* published an investigative report titled "40% of residents living near Realgar Ore Mine poisoned by arsenic in Hunan: 7 members of a family died of cancers". The trends analysis published by Sina Weibo shows that the index of discussions on the topic "arsenic poisoning" increased from only 77 on the previous day to 7,419 on April 13, 2014. The in-house data also suggests the investigative report at *Law Evening News* is the most circulated content about this topic by these Weibo tweets on that day. Although online attention to this topic quickly dropped the next day (the index was 1,586), there are two more minor peaks on April 16th (when *Netease* launched a photo collection named "poisoned village" on the same topic) and April 18, 2014 (when *Economic Reference Daily* published an investigative report further revealing the arsenic pollution and the suffering of cancer victims in a wider area near the mine). The Netease photo collection attracted 40,160 comments from Internet users. These two newspaper investigative reports and *Netease*'s photo collection have built up dramatic visual symbols to engage Internet users. The dreadful and terrifying images of the destroyed environment and the miseries of local residents construct visual symbols of arsenic poisoning and pollution and cancer villages either via textual descriptions or pictures. These visual symbols engage and shock Internet users.

Of importance here is the feeding back of investigative reports onto online environmental discussions, which enables the continuity of these discussions on certain environmental agendas. In this case, another important impact of these investigative reports was to maintain the continuity of online attention to the topic of cancer villages. One can see the popularity of the topic of arsenic poisoning tracks the popularity of the topic of cancer villages on Sina Weibo. Both see two peaks in the overall trends in popularity of the term "arsenic poisoning" and "cancer villages" appearing on Sina Weibo April 16, 2014 and April 18, 2014.

However, compared to entertainment topics, environmental problems do not enjoy much priority in gaining online attention. For example, the topic of arsenic poisoning obtained a maximum index of 7,419 on April 13, 2014, while a scandal about an affair by a celebrity, Wen Zhang, can capture much more intense online attention, with an index of 400,162 on March 31, 2014. In this sense Internet users prefer entertainment topics to serious environmental topics. This, nevertheless, is not a problem with investigative journalism, but with the interests and aesthetic of Internet users themselves.

5.5   Conclusion

Online environmental movements and investigative journalism play their respective roles in two separate spaces, though influencing each other from time to time. Online environmental crusades that are discussion-centred have captured the attention of and engage offline investigative journalism. Investigative journalism has shown a considerable interest in the heated discussions of environmental problems taking place on the Internet. When these online crusades enter into the coverage of investigative reports they appear in the public and political arena which is traditionally dominated by mainstream news media. At the same time, investigative journalism tries to retain its independence from the influence of online environmental movements, protect its agendas and avoid being dominated by the topics discussed online.

If examining the longitudinal development of environmental agendas – either those concerned with nature or with well-being – within the past decade, one finds there is continuity in the development of environmental agendas. The contribution of online activism to this development has already been appropriated and incorporated by offline investigative journalists into their conventional practice. Investigative journalism has a new function, namely to check the credibility of fragmented online information and reveal a complete picture of events. In doing so investigative journalism enlarges civil discourse about environmental protection, opposing the dominant discourse that prefers economic growth.

The publication of investigative reports engages and mobilises Internet users to a certain extent. However, the interests of online communities actually are driven by many factors including the existence of the tabloidization tendency of online media, preferring entertainment to serious topics. Therefore it is difficult to say investigative journalism is the only force that effectively mobilises the public. It is more accurate to take the joint power of online environmental crusades and offline investigative journalism into account. The two work in parallel, developing in their own spaces but attracting the attention of each other from time to time. The contribution of both makes the continuity of certain environmental agendas possible in contemporary China, exercising counter-power against the hegemony of modernisation. However, this power is potentially limited by the nature of the main interests of the public on the Internet, which tend to be apolitical and entertainment-related (Lum 2006; Leibold 2011; CNNIC 2014).

6
Hegemony and Counter-Hegemony: Investigative Journalism between Modernisation and Environmental Problems

Modernisation, a hegemonic ideology in Chinese society, justifies the state's policies for development. It has become a paramount guideline that is even beyond the control of the state. The state has already realised the threat environmental risks pose to its rule and thus attempted to promote an ecological modernisation (Zhang, Mol et al. 2007; Huan 2007; Lia and Langa 2010; Mol 2006; Yee, Lo et al. 2013). In spite of that, the state cannot resist the lure of modernisation, spurred by its endless desire for development. As modernisation has become an effective disguise for ruling interest groups to make profits for their own sakes (Liu 2007), the state, under the strong influence of the representatives of these ruling interest groups (Cai 2014), is unlikely to slow down the pace of modernisation for the benefit of the environment. Environmental problems in fact result from the headlong pursuit of profits by these groups, enabled by state capitalism (Liu 2007). Therefore the hegemony of modernisation by its very nature is the hegemony of state capitalism and of interest groups rather than the hegemony of the state.

Clearly, investigative journalism has constructed a discourse of environmental risk that is in contrast to the discourse of modernisation encouraged by Chinese governments for over half a century. Through embedding Marxist environmentalism in the discourse of environmental risk, investigative journalism exposes the true nature of modernisation on the one hand and on the other reveals the connection between social and environmental injustice and inequality. In doing so, investigative journalism echoes both the neo-leftist and rightist critiques of capitalism and authoritarianism respectively. This construction of environmental risk discourse further uncovers a deep crack in China's modern life, resulting from the disconnection between the past and the present caused by the hasty transformation from an agricultural to an industrial society. The bifurcation between the two discourses reflects the rift between environment and modernisation as well as the divergence between people and the state. Thus, the contrast of the two discourses is a mismatch between the hegemony of modernisation and the reflection of ordinary people on their daily lives in the present industrial society, underlying which is the logic of state capitalism by its very nature. This contrast is exaggerated by intense domestic social problems and tensions. By constructing this discourse, investigative journalism demonstrates a counter-hegemonic force to a certain extent. A pressing question for us to think about, however, is to what extent the force can be counter-hegemonic. Answering this question requires us to examine the limitations on the practice of investigative journalism and the scale and scope of the influence of investigative reporting within the wider social and media context of China.

6.1   The hegemony of modernisation

In Gramsci's perspective, the concept of hegemony is used in relation to how power works. In other words, the hegemonic groups seek the spontaneous consent of "the governed" in order to achieve hegemony (Gramsci 1971: 527). Thus consent plays a crucial role in exercising hegemony (Gramsci 1977). In China, rapid economic growth is a persuasive target set by the hegemonic ruling groups for the subordinate groups to achieve. The market is at the service of the state and vice versa. The target of achieving economic modernisation is the state's means to exercise hegemony effectively in Chinese society. Modernisation has been accepted by the public as being the most important and urgent thing to do. This consent helps the CCP justify its rule and its economic policies. The Chinese people's dream of modernisation underlies the CCP's ruling legitimacy, since it is the CCP that will lead China down the road of modernisation, becoming a wealthy and strong country.

Starting from Mao's time, it has been agreed from the top to the bottom that the whole society should fully pursue economic modernisation. This task has been constructed as a primary political ideology, beyond its economic meaning. The occurrence of a series of social and political events and even traumas, such as the Great Leap Forward, are all inevitable side effects of, or at least highly associated with, this ideological attempt to modernise China. More recently, following the failure of pro-democracy movements in 1989, China's top leaders have endeavoured to shift the attention of the general public from political reforms to reforms in the economic arena. The aim of achieving economic development has replaced many other tasks, becoming a prioritised philosophy that supports the rule of the state. Deng Xiaoping's "cat theory" not only legitimises becoming rich but also justifies the success of the regime by judging it on its economic achievements. Consumerism, for example, has successfully shifted the attention of ordinary people from political reforms to materialist enjoyment for much of the time.

The pursuit of modernisation and economic achievements has not only united the Chinese people to march towards the same goal under the leadership of the CCP, but also conceals the social and political problems China is facing, such as social inequalities, the lack of political reforms and the absence of legal transparency. The priority for economic growth thus makes people overlook other basic needs of a healthy society, such as democratic and political needs. The idea is that while people are celebrating their advanced material achievements in modern China, not many of them will truly care about what China needs in order to become a better country. As a hegemonic ideology, modernisation plays its part well in consolidating the CCP's rule and continuing China's dream of a harmonious and prosperous society. By promoting modernisation, the party-state thus achieves its purpose of replacing political reforms with economic reforms.

Investigative reporting on environmental problems is therefore not merely about these problems but are actually about modernisation, revealing the true face of a marketised economy under authoritarian rule. This is largely because there is an intimate and inseparable relationship between modernisation and the environment. The environment is central to ordinary people's daily lives and a key indicator for whether or not they can live a happy, healthy and safe life. Environmental problems are not merely problems about the environment but problems about society and modernisation. In a society that is currently full of intensified social conflicts and tensions, environmental problems could become the fuse igniting the accumulated social grievances. The discourse of environmental problems is that of modernisation –in particular, the discourse of the consequences of modernisation. This discourse harmonises with various views on the economic reforms in Chinese society such as those of neo-leftists and of rightists.

6.2 Revealing environmental problems: environmental risks and shock

The revelation of environmental problems through the coverage of investigative reports lies in constructing both environmental risk and the shock China's industrialisation has brought to ordinary people in relation to their living environment. The hegemony of modernisation is offset by the warning bell of environmental risk and the broken link between the past and the present.

Terrible pollution, dying vegetation, animals and cancer victims, riverbanks torn by sand-digging, rivers disappearing as a result of hydroelectric plants and dams, the frequent occurrence of landslides, droughts and earthquakes, expanding deserts, exhausted natural resources and collapsing ground – all of these paint a miserable picture of a deteriorated environment and a dim future for the Chinese people regarding their unhealthy and risky living environment. None of these environmental problems can be solved immediately and efficiently. No technologies are available to permanently eliminate any of these environmental problems, though perhaps some of them can be mitigated under appropriate governmental administration. Many of these environmental problems are already beyond the ability of the government and science to handle. Even so, Chinese governments continue to advance and prioritise economic growth over environmental safety.

The environmental risks exposed in investigative reports are beyond the capability of Chinese society to manage, and therefore the social and technological rationality that is essential to modernisation is open to question. Modernisation and its Promethean environmentalism centre on social and technological rationality which places confidence in governmental administration, elites and the application of technology to fix problems emergent in the process of industrialisation and modernisation. However, what investigative reports have revealed are problems that cannot be solved by any of the current technologies the government possesses. This is an important reason why investigative reports are critical to modernisation. The invasion of smog, the presence in many places of poisoned soil and polluted water, the threat of cancer, the paramount power of natural disasters such as drought, flooding, landslides and earthquakes – all of these signal the arrival of a risk society where governments find themselves incapable of dealing with pressing environmental problems. This is a picture of potential doom to humanity: what has been done for the sake of modernisation will bring insoluble consequences for the environment in return. This reflects the "prophetic function" of journalism (Neuzil 2008).

The construction of environmental risks is based on the experience of investigative journalists and ordinary people, reflecting the importance of individuals rather than social organisations in the construction process. It may be the stories of ordinary environmental victims' miserable lives or the first-hand accounts and horrific images presented by investigative journalists – they all offer vivid pictures, enabling readers to share the experiences and resonate with them based on their own experience. Not only investigative journalists but also the news sources themselves and most of the readers, if they were born before the new century – especially for those who were born by the 1980s – are the generations that have gone through the transformation from an agricultural to an industrial society. They still have some memory of the harmony between nature and humanity from childhood. The conversations with investigative journalists in the interviews, for example, often involve their descriptions and comparisons of what their hometowns looked like when they were children and what they look like now. This nostalgia penetrates investigative reports, reflecting their longing for the harmonious relationship between nature and human beings as discussed in Chapter 4. The individual experiences of both news sources and investigative journalists echo and connect with what ordinary people express on the Internet about their own experiences and memories of the past and comparison with the present. Underlying their yearning for the past and loathing of the present is the shock felt by people, since they do not know what to do when facing the dramatic transformation brought by industrial society as a result of the wrecked link between the past and the present. Investigative reports on environmental risks construct this kind of shock. What environmental investigative journalism has done hence includes reflecting this sense shock and discontinuity between an agricultural and an industrial society. Investigative reports on environmental problems provide a source of nostalgia for readers.

As discussed above, on the surface, environmental problems are simply problems with the environment. However, by their very nature and as constructed by investigative journalism, environmental problems are more than that. Instead they are also problems with society. This is because the psychological shock felt by people as a result of environmental problems is related not only to people's concerns over their safety and health, but also to their nostalgia for "the golden past" in terms of the environment and the agricultural society in which most generations of the general public currently alive grew up. The hasty development of Chinese society has left no time for people to bid a proper farewell to their original pre-industrial agricultural society and to reflect on the changes in their living environment and lives. Instead, all of a sudden, ordinary people have been dragged into the wave of industrialisation and urbanisation and forced to face the risks presented by a society that they and their ancestors have never experienced before. One more important reason is that when an agricultural society is suddenly turned into an industrial society, it is very difficult for the majority of the Chinese population, which was an agricultural population living in rural areas (Jing 2010), to automatically become a non-agricultural population in many aspects such as social benefits, employment, residential status (*hukou*) and customs. In this sense, environmental problems are indeed a dimension of the broken connection between pre-industrial society and modern industrial society, which is the consequence of rapid modernisation, driven by the potent hegemony of capitalism.

Therefore, the construction of environmental risks and the shock felt as a result of environmental problems opposes modernisation in two ways. First, through implying that pressing environmental problems are something that cannot be dealt with effectively by using current technologies, the construction conveys a message that modernisation is not as beneficial as the party-state has boasted that it is. The centrality of technologies and the confidence in them is actually seen as somewhat false. Modernisation may be leading people to the abyss of environmental catastrophe from which no one – especially the underprivileged – can escape. The fact that the government and officials carry on doing what they believe is good for modernisation, as revealed in investigative reports, leads to an interpretation that environmental risks are occurring as a result of governments' ignorance. Investigative reports are critical of both the market and the state, without holding out a slice of naive hope that the state can resolve these problems.

Second, environmental problems reflect the rupture in Chinese society and people's discomfort with the changes that have occurred as a result of fast modernisation that has driven society to shift from a native agricultural to a modern industrial society. The shift has happened so fast that ordinary people have not adapted to the changes emergent in the process of transformation. Especially now, when they realise they are facing environmental problems, they, including investigative journalists themselves, have started missing the past. Such nostalgia, feeling a sense of what has been lost through the modernisation process and worrying about the future characterise both investigative reports and online discussions. Both aspects – environmental risks and shock – are two important negative consequences of modernisation that the party-state cannot avoid facing in the end.

Apart from these two aspects, environmental investigative reporting has revealed the true face of modernisation, further breaking down the sweet promises the CCP has made to its people.

6.3   Exposing the true nature of modernisation

In the face of the hegemony of modernisation, another important thing investigative journalism on environmental problems has done is to reveal the true nature of modernisation by constructing a position that shows the rivalry between the privileged and the unprivileged, as well as the collaboration between political and economic forces. The privileged are often portrayed as environmental problem generators, whereas the unprivileged are environmental victims. In doing so, investigative journalism breaks down the myth of modernisation and the sweet dream of a "better life" woven by the myth, since a "better life" may only exist for the privileged. As discussed already in previous chapters, environmental investigative reports show how problems occur generally as a result of the economic activities of privileged individuals or institutions. These individuals and institutions often possess economic and political resources and are major actors in local political and economic activities. Their joint possession of economic and political resources implies collaboration between political and economic elites, such as governments, officials and entrepreneurs, which veils the true face of modernisation.

The true nature of China's modernisation is highly associated with the needs of powerful international and domestic interest groups pursuing maximal profits. In this sense the state has been kidnapped by interest groups and capital in two main ways. First, the major industrial projects that characterise China's modernisation are controlled by ruling interest groups that enjoy both political and economic capital in China and therefore are able to benefit most from these projects, ranging from national ones, such as the construction of the Three Gorges Dam, to local ones, such as the setting up of PX projects in local areas. For a small group that is part of the ruling class, modernisation has become their major source of making money (Ho 2013). Here interest groups refer to those groups whose members share common interests and command resources that complement and fulfil one another's needs. Interest groups that benefit most from modernisation are those that enjoy easy access to political resources and can influence policies and policymaking, such as the princelings who are descendants of influential high-ranking

CCP officials at the central level, officials at local levels, their relatives, friends and allies in the economic field (Muldavin 1998; Ho 2012, 2013). These princelings and local officials sometimes are themselves important figures in the economic field (Li 2009; Ho 2013).

In China, authoritarianism neither allows the general public much flexibility in accessing political and economic resources, nor has enough political constraints on the influence of interest groups (Horowitza and Marshb 2002; Steinberg and Shih 2012). Political power is limited to a small group of political elites, especially to princelings and their associated interest groups, excluding the general public from power (Xiang 2012; Bo 2008; Li 2009). Economic resources are carefully allocated according to the hierarchy of the administrative system. Those at the top of the ladder of the social hierarchy usually enjoy more political and economic resources than those at the bottom of the ladder. After the 1980s' economic reforms, access to economic resources was opened up to the general public. However, those who enjoy easy access to political resources quickly gained more opportunities to access economic resources. This has led to the concentration of economic resources in the hands of a few political elites and the prevalence of clientism in the country (Oi 1985; Ho 2013). Ordinary people are left behind at a spiralling speed. The confluence of political and economic resources results in a situation where those who are included in the gentlemen's club of interest groups benefit most whereas those who are excluded benefit less or least from the economic reform. In this sense, big policy decisions and decision-making by the state are beneficial to the interest groups who are central to the decision-making process. The personal networks (*guanxi*) in Chinese society are complex and intertwined. It is difficult to separate these interest groups one from another. A true picture of these interest groups would show that they are interconnected in one way or another and comprise a huge set of networks that manipulate all kinds of issues in society. In this way the state has been kidnapped by interest groups.

A cruel truth about modernisation in relation to environmental problems is that the party-state would only be willing to deal with those specific environmental problems such as smog that all Chinese people suffer, but will leave other environmental problems behind. Only when interest groups want to re-allocate their interests and profits will specific environmental problems be allowed to emerge in political agendas and be handled.

Second, the state is kidnapped by and equivalent to capital when governments at all administrative levels are lured by economic profits and driven by the logic of state capitalism. As already discussed in Chapter 1, policies have been made which facilitate the inflow of capital from within and outside China. Governments promulgate favourable tax and land policies and even lower environmental barriers in order to attract national and transnational corporations. Governments are on the side of enterprises in terms of making profits. The more profits enterprises make, the more taxation and political achievements local governments gain. Therefore, governments are the accomplices or even the core of interest groups in many cases. In other words, the state is capital, while capital is only a strategy the state has been using to fulfil its purpose of achieving the legitimacy of its rule. Under such circumstances, even if the state has realised the risks posed by environmental problems and would like to relieve the problems by promoting an ecological modernisation, the willingness of the state may be offset or overcome by the enthusiasm of local governments for economic growth (Lia and Langa 2010) on the one hand and, on the other, ecological modernisation still prioritises economic growth, underlying which is the logic of state capitalism.

Take the construction of the Three Gorges Dam, for example. The Three Gorges Dam took 18 years to complete, and its construction involved an investment of around RMB 20,000 billion.[1] Why did China insist on going ahead with the project? The official reasons the Chinese government has given of course are reasonable, including anti-flooding and the resolution of electricity shortages (Shapiro 2012). But the real and most important reason behind the construction is perhaps the economic reason. This is not only because the construction itself generated numerous jobs and opportunities for building companies, but also because of the huge income from electricity and transportation. One of the biggest economic benefits of its construction is the generation of electricity. The capacity of hydro-electric power in the Three Gorges Dam is 18.20 million kW, generating 8,470 billion kWh electricity annually. The Three Gorges Dam's electricity transmission system covers more than 160 administrative areas at the county level in the East, South, Middle and Southwest regions in China.[2] The construction of the dam, therefore, is closely linked to, and benefits most, China's electricity industry that is believed to be under the control of the family of Li Peng (Oakes 2004). Since the start of 2014 there have been widespread accusations of corruption by Li Peng's family, who have made profits from the Three Gorges Project. The corruption not merely refers to his family's monopoly of the electricity industry of China, but also Li Peng's influence in the bidding process which decided the selection of the building companies for constructing the dam, as revealed by investigative reports at *Times Weekly* (*shidai zhoubao*) in 2014.

However, most investigative reports on environmental problems are unable to touch such issues surrounding the dominant interest groups at the central level, such as the business affairs of the princelings. When reporting on environmental problems associated with the construction of the Three Gorges Dam, for example, the focus of investigative reports has been placed on the environmental problems appearing in the regions of the Yangtze River as well as probing whether the construction of the dam should be held responsible for environmental problems, such as droughts, and disasters such as earthquakes and landslides. None of these investigative reports has dared to associate the environmental problems with the monopoly of the family of Li Peng. The wrongdoings of the Three Gorges Group, which was under the control of the Li Peng family, were only revealed in the coverage of investigative reports on corruption (rather than environmental problems) involved in the construction of the dam (rather than environmental problems) by *Times Weekly* in 2014, when Xi Jinping, the current president, declared a war on corruption and "tigers", and removed Zhou Yongkang, the boss of China's petroleum industry, from power and put him behind bars. Domestic and overseas observers believe the next anti-corruption target of Xi could be Li Peng. Except under such political circumstances, no investigative journalist would dare to touch the princelings and interest groups at the central level. This is a major limitation on the influence of investigative reporting.

What investigative journalists reporting on environmental problems do dare to touch is the monopoly and collaboration of interest groups at local levels. The investigative report: "Investigation into the wrongdoings involved in the Jingshajiang Hydroelectric Project" *(jinshajiang shuidianzhan weigui diaocha)* in *Southern Weekend* (June 18, 2009), for example, directly revealed the misconduct of interest groups in constructing the Jiangshajiang Hydroelectric Project. In this case the interest groups are mainly local governments, hydroelectric development companies, banks and investment companies. The intertwined interests of these actors pushed them to circumvent the governance of the central environmental protection ministry and start to construct hydroelectric projects without passing the environmental evaluation of the environmental department, which has caused irreversible damage to the local environment in the region of the Jingshajiang River. It is still better than nothing, as such a revelation can at least expose the true face of modernisation and let the general public know it is not what it claims to be.

6.4   The connection between environmental problems and social injustice

The true face of modernisation is further unmasked when investigative reports on environmental problems link environmental problems to social injustice. Inequalities and injustice indeed exacerbate environmental degradation for its victims (Shapiro 2012). Investigative reports clearly uncover the causal relationship between environmental problems and the economic activities of privileged groups – especially those engaged in business and their allies in government agencies – and this reveals that state capitalism is the real hegemonic force behind modernisation and that interest groups play a crucial part in advancing modernisation and causing environmental problems in China. This in turn leads to the construction of a connection between environmental and social injustice in environmental investigative reporting which further exposes the downside of modernisation and has the potential to overthrow the hegemonic status of modernisation. Modernisation, like many other types of development in human societies, leads to and widens (rather than bridges) the gap between social groups and individuals who have good life chances and those whose life chances are constrained by the limited resources they possess. The former enjoy the fruits of modernisation, while the latter bitterly suffer the environmental costs incurred in the process. The establishment of this link tends to aggravate social grievances that are already intense.

Whether individuals and social groups are judged as environmental problem generators or victims depends on their relationship to environmental problems. Environmental problem generators are usually those who exploit local resources to achieve economic profits and are able to leave the localities they have exploited and never even lived in behind. However, environmental victims are often those who live in these despoiled locations and are unable to leave to live in a cleaner, safer environment. Resource exploiters are those who are positioned at the top end of capitalist production and who are able to make use of their already considerable resources to exploit more resources, to make more profits. Prominent among them are entrepreneurs, merchants and officials. Environmental victims, however, often sit at the foot of the social and economic hierarchy, on the receiving end of the consequences of economic activities and decisions made by resource exploiters and unable to leave the environment that has been destroyed in the process. They include people from the bottom of society, such as peasants, workers, farmers, fishermen, villagers and herdsmen. They may also include the majority of people in underdeveloped regions, such as the west and middle regions of China, as well as animal species and nature. According to Foster (Foster 2000, 2010; Foster, Clark et al. 2010), a healthy environment should have an equilibrium between resource usage and reproduction, that is, after being used and consumed in an environment resources will be recycled and returned to that environment. However, the current mode of capitalist production and the global logistic system has broken down this equilibrium. Resource exploiters resource unsustainably, take products away and leave the environment impaired by their exploitation. Environmental victims, who often cannot leave, even lack the opportunity to achieve the environmental justice they seek. For example, by 2006, one seventh of the underground area in Shanxi Province was hollow and around 2.2 million villagers living in 1,900 villages became "ecological refugees" who, however, were not able to leave their homes. The coal-mine owners whose activities were responsible for the disaster, however, left the problematic land (Zhu and Long 2012). In this way the discourse of environmental problems is extended to include the discourse of environmental injustice, an extension of existing social injustice in Chinese society. Even when facing some general environmental problems, such as smog, resource exploiters still enjoy an advantage over other environmental victims. The former are able to purchase air filters, air masks or even the latest inventions – "air cans" (*kongqiguantou*) for the sake of their own health, or leave China for other countries. Some of them have never even lived in China.

The connection between environmental problems and social injustice that is constructed in investigative reports leads to the ultimate question as to what modernisation will truly bring for us. This question and associated doubts pose a critical challenge to, and prompt urgent reflection on, the dominant discourse of modernisation.

This discourse first of all reveals that modernisation is not a utopian ideal for everyone on an equal basis. China has integrated its orthodox socialist utopian ideal into the target and promotion of modernisation. Social equality and harmony is the ultimate goal of socialism and thus socialist modernisation. Those promoting modernisation in China seemingly adhere to a critique of Western capitalism and advance a utopian ideal of achieving national prosperity (*minzufuqiang*) and building an equal and harmonious world (*tianxiadatong*) for its people. This ideology of modernisation stresses the importance of a centralised administration, an advanced market and high technology (Dong and Yang 2007; Zhang 2007). By contrast, investigative reporting on environmental problems reveals that modernisation does not necessarily lead to social equality and national prosperity. Instead, it actually broadens the gap between individuals and social groups: that is, the rich become richer and the poor poorer. It is a small group of individuals and social groups that manipulate and benefit from modernisation, leaving ordinary people behind suffering the environmental costs of modernisation. Furthermore, neither centralised administration nor advanced technologies are able to resolve the environmental problems that have emerged in the modernisation process. This exposure turns the ideology of modernisation upside down and overthrows its utopian ideals. Therefore the discourse of environmental injustice poses a critical threat to the discourse of modernisation promoted by the regime.

This thus raises a question mark over the philosophy of seeing development and marketisation as a remedy to cure poverty and a guarantee for a happy life for all which underpins the ideal of modernisation. One important purpose of modernisation is to erase poverty in China, with the promise of leading the people to a happy life. However, this seems to collapse when investigative reports reveal the fact that environmental victims are left destitute because of the loss of life opportunities and health. The precedence given to wealth over health is proved to be wrong. Environmental victims, such as local residents in regions where natural resources have been exhausted, fishermen, herdsmen and cancer victims in cancer villages endure poverty because of the loss of jobs and income and illness induced by environmental problems. These people who are abandoned by modernisation are facing poverty, disease and premature death.

In addition, being the "sick man of East Asia" (*dongya bingfu*) was a source of deep shame for all Chinese people and has been seen as the stereotyped image of China since the 20th century. One important purpose of modernisation is therefore to change the image of being the "sick man of East Asia". However, when China is now being portrayed as a country full of cancer victims and patients suffering high levels of lead in their blood and respiratory diseases, the image of being the "sick man of East Asia" has returned in a different form. Modernisation no longer holds the same promise to free the Chinese people from poverty and diseases. Therefore the sweet dream of modernisation delivering wealth and strength is overturned.

In this sense, the two ways of exposing the true face of modernisation, discussed above, resonate with the neo-leftist condemnation of capitalism as well as the rightist criticism of the state's authoritarian rule.

6.5   The rift between two discourses

The contrast between the two pervasive discourses regarding the environment and modernisation coexisting in Chinese society at the present time is outstanding. The official discourse of modernisation, promoted ever since 1949 among the people by the state, is no doubt the nationally prioritised task that Chinese society aims to fulfil. The discourse of modernisation has been woven intensively into economic policies, official speeches and the coverage of Party organs. Particularly for local governments, the growth of GDP and economic development in their regions has become their clear target, and the criteria for judging their political achievements as well. When environmental problems and issues have come to the attention of the central government, the state has started to advance environmental protection and so-called ecological modernisation, although still sticking to its priority for economic growth. Despite this, the discourse of modernisation remains dominant in China's official discourse.

The discourse of modernisation embodies the economic logic of modern Chinese society that has rapidly shifted from an agricultural to an industrial society. Economic development and an increase in GDP has been the top priority for Chinese governments and even for the whole society. GDP growth and other economic indicators, such as the rank of the economy in the world, are the criteria for measuring the level of modernisation of Chinese society as well as the degree to which how the CCP's governance is successful. In 2011, for example, China's overtaking of Japan as the world's second-largest economy marked the key achievement of modernisation. How far the country is from modernisation is a continuously pressing question for China. The discourse of modernisation encourages and justifies a wide range of social and economic activities, such as the building of dams and hydroelectric projects, mining, the launching of PX projects and so on, no matter what kind of environmental costs and consequences these activities may have.

The discourse of modernisation offers a set of values and criteria for evaluating the changes in society and the pros and cons of the state's policies. This is a discourse that is able to nurture ordinary people's material satisfaction. This discourse encompasses a social and technological rationality that is believed to be able to overcome any risks triggered by human activities. In other words, this is a Promethean discourse of the relationship between the environment and development. While saying that, this discourse of course also enables the ruling CCP to consolidate its legitimate rule, since the rule of the Party has successfully led to success in economic growth, which is believed to offer opportunities for people to live happy lives.

In contrast, the coverage of environmental investigative reports contributes to the rise of a strong discourse of environmental risks in society. The discourse of environmental risks portrays a challenging but depressing picture of the problems encountered in the natural environment, and predicts extraordinary environmental risks for the Chinese people in the near future. This discourse is warning of the environmental impacts of modernisation, reflecting on our current status, pining for the past and predicting the far from bright future.

This discourse arises from fear of the devastating power of nature and the inescapability of environmental risks as a result of human economic activities. This discourse also reflects the anxiety of Chinese people over entering an industrial society. The direct and indirect personal experiences and the emergence of tragic stories of environmental victims naturally drive ordinary people to reflect on their current deteriorated environment and possible problems to their health and safety caused by it. This type of discourse constructs a different value system and standards, such as environmental justice and safety, with which to judge the level of happiness of people's lives. It reflects people's dissatisfaction with their living environment and concern over their future, which presents a big challenge to and casts doubt on the social and technological rationality that justifies and underlies the discourse of modernisation. Moreover, this discourse is characterised by a feeling of doom, the environment's punishment for Chinese madness in modernisation. Environmental risks, moreover, are seen as beyond the ability of any governments to solve. Therefore, the discourse of environmental risks challenges the legitimacy of the ruling Party, since it claims to work for the happiness of the Chinese people, something which nevertheless turns out to be unrealisable in a risk society. As a result of this, Chinese society seems to be engaged in two extremely different thought processes at the same time, and the conflict between them seems impossible to resolve.

The rift in these two pervasive discourses is between the environment and modernisation, reflecting the conflict between the people and the state and a big crack in present society caused by the rapid transformation from an agricultural country to an industrial one. Modernisation has an intrinsical conflict with the health of the environment. As long as economic growth and development remain the top priority for China, the damage to the environment will not end and the environment will inevitably be sacrificed to this priority, no matter what advanced technologies Chinese governments adopt. Driven by this Promethean attitude, confidence in the ability of technologies and administration to fix any problems emergent in the process leads to a common pattern in resolving environmental problems: carrying out environmental governance after the occurrence of environmental deterioration. However, with the proliferation of environmental problems, Chinese society is realising that the risks associated with environmental problems are beyond the capability of any single government. The priority placed on modernisation and capital accumulation requires the discourse of modernisation to take precedence over any other discourses, while the environment stands to benefit from the discourse of environmental risks which contribute to drawing the attention of the public and governments to environmental protection.

The risk discourse does present challenges to the existing power structure in Chinese society and raises the potential for change. At the same time, however, it contributes to the state's control as it increases knowledge of the environment and prompts individuals and institutions to discipline themselves.

6.6 Environmental investigative journalism forging a counter-hegemonic force

The hegemony and counter-hegemony theories of Gramsci offer us an excellent conceptual framework to understand the role of investigative journalism in opposing the hegemony of modernisation in China. This chapter argues that investigative reports, through their ability to construct and convey the symbolic reality of environmental problems continually and in detail to their readers over the past two decades, has cultivated a counter-hegemonic discourse against the dominant discourse of economic growth. There are four reasons for this.

Environmental investigative journalism is counter-hegemonic first because investigative reports raise an environmental critique of modernisation, revealing the true face of modernisation and destroying the myth and the dream that modernisation has woven for the Chinese people. Investigative reports on environmental problems portray the incompatibility between capitalist modernisation and a healthy and sustainable environment. Environmental deterioration is depicted as a consequence of modernisation driven by the logic of capitalist development. Environmental risks are brought before the eyes of the public and society is warned of potential environmental catastrophes.

Second, investigative reports on environmental problems show their counter-hegemonic power through the exposure of the ways in which advantaged individuals and interest groups exploit natural resources and the environment, while the disadvantaged are deprived of life chances and marginalised during the process of modernisation, and suffer most from the environmental destruction. The incompatibility between ecology and economy indeed reinforces the rift between the advantaged and disadvantaged. In doing so, investigative reports are vehicles for the voices of resistance and opposition, mirroring the suffering and fears of ordinary people – especially the disadvantaged – for the deteriorated environment and their diminished life chances and marginalised social, political and economic status.

Third, the counter-hegemonic force of investigative reports also lies in the challenge posed to the supremacy of technologies that is central to modernisation. These investigative reports suggest that advanced technologies are not a cure-all for every environmental disease and even the application of technology itself, such as in the building of dams, can lead to environmental destruction. This is another aspect in which investigative reports break down the myth of modernisation.

Last but not least, the importance of investigative reports lies in keeping environmental agendas at the forefront of the public's mind, which helps establish environmental civil society's "war of position" (Gramsci 1977) through creating a counter-hegemonic force against the existing hegemony of modernisation. Although the concept of civil society in China suffers several flaws, as discussed in the Introduction, the emergence of civil rights movements, including environmental movements, at least generates – perhaps small but still promising – spikes of resistance.

This function of investigative journalism originates from the symbiotic relationship between the state and investigative journalism. Chinese investigative journalism has been historically expected to function as an "ideological state apparatus" by the state. The state that has realised the challenge presented by environmental problems to its rule needs investigative journalism to expose them, though not expecting this to be counter-hegemonic to modernisation. On the other hand, investigative journalism requires a supportive attitude on the part of the state in order to achieve reasonable autonomy and space for its own development. This is why environmental problems are often seen as less politically sensitive topics if compared to socio-political issues.

However, such a counter-hegemonic force may be limited by four factors. The first is the restrictions placed on the practice of investigative journalism. For example, as discussed already, investigative reports do not touch the involvement of princelings in advancing economic projects that are responsible for environmental problems. They only can criticise the collaboration between political and economic elites and governmental clientism at local levels, though these still exemplify well the whole situation in China and tear a small hole in the curtain covering the whole scenario from which we can understand what lies behind economic development and environmental problems.

These restrictions, nevertheless, vary from time to time, due to China's complex media–state relationship. Different levels of media control are often seen along with diverse requirements by the party-state during various periods of time. The boundaries of reporting limitations thus change over time. The current harsh political and economic conditions of the news media do not offer a good environment for investigative journalism. If the tightened media control in the Xi era remains and news media become more and more dependent on governments for economic returns, investigative journalism will have little space and autonomy for its practice. Therefore, the golden era of environmental investigative journalism in the 21st century may disappear one day. Nevertheless, although the increased tightening of media control since 2013 does not present a positive scenario for the development of environmental investigative journalism, it does not mean the severe winter for investigative journalism will last forever. Environmental investigative journalism needs to wait for the right time as well as to develop smart tactics to deal with the limitations. However, these ebbs and flows in environmental investigative journalism practice inevitably undermine its counter-hegemonic power.

The second is the influence of investigative reports. Measuring the power of the counter-hegemonic force embodied in the discourse of environmental risk has to take the scale of these reports' influence into account. The number of environmental investigative reports is so small that they are only a drop in the ocean if compared to the huge volume of other news reports, such as hard news. Therefore, if considering the number alone, the influence of environmental investigative reports would appear to be overwhelmed by a huge number of hard news items – especially those in the coverage of party organs – praising the performance and consequences of modernisation. However, in terms of influence we cannot merely consider the number of reports. Instead the quality and credibility of reports is more important. Of course, studies examining readers' reception of environmental investigative reports and "normal" hard news are required in order to understand these issues precisely, which is beyond the scope of the present book. In spite of that, one thing that is for sure is that news is unlikely to be seen as credible if it downplays environmental deterioration and therefore does not fit the reality readers see every day.

The third is the general outlook and interests of readers, which are evolving and elusive. Readers' outlook and interests have become progressively more important for understanding the influence of environmental investigative reports. The attention of the public is essential both for the public mobilisation of investigative journalism and for feedback from investigative reports to online discussions. The influence of investigative journalism cannot be taken for granted without considering the world view of readers who sit at the receiving end of environmental communication. Along with the tabloidisation of the Internet, the apolitical world view of readers and the prevalence of consumerism may drive readers to pay attention to amusing entertainment topics and their material enjoyment, meanwhile becoming uninterested in serious environmental issues. If it fails to attract extensive public attention, environmental investigative journalism may lose its position in Chinese news media outlets that have to operate like businesses in the market and therefore must pursue profits. This may mean that environmental investigative journalism cannot be sustained.

The last, but not the least, factor is the level of civil society development in China. Given the interaction between the two, the more advanced civil society becomes, the more influence environmental investigative reports can have. As discussed already, nevertheless, China's civil society itself is constrained by the existing state–society relation in the authoritarian context, lacking the capacity to democratise society. The close relation of civil society to the state as well as the constitutive role of civil society in maintaining the state's legitimacy leads to a restricted function of civil society. The hope of intensifying the counter-hegemonic force of environmental investigative reports thus is given to the possibilities for changes – no matter if trivial or major – opened up in the rise of a civil society with Chinese characteristics.

Notes

Introduction

1. Xi made a speech in the conference in which he stressed on the extreme importance of ideological work and of "propaganda struggle" (*yulundouzhen*). Since
   the conference, the ideas of "propaganda struggle" or of "ideological struggle"
   are prevailing in Party organs' coverage and local officials' discourses. This is
   seen as the backward development of speech freedom in China.
2. http://news.xinhuanet.com/newmedia/2013-03/29/c_124520703.htm
3. These reasons arise from the author's interviews with environmental investigative journalists from 2011–2013.
4. All names of interviewees are anonymous. Therefore, pseudonyms are used
   instead. In the book, the surnames of journalists and activists appear before
   their first names.
5. The Cultural Revolution was launched by Mao Zedong from 1966–1976.
6. Chengguan is short for a functional department of Chinese government
   called City Urban Administrative and Law Enforcement Bureau. Usually in
   media content or daily conversation, "chengguan" means staff members hired
   by the Bureau to enforce the law and urban regulations.
7. In recent years China has seen events in which patients killed doctors because
   of disputes. Usually they regarded doctors as having treated them inappropriately and in an unfair way. A recent and influential case happened in Wenling
   City, Zhejiang Province, on October 25, 2013. In this case, a patient who
   had an operation on his nose killed a doctor and severely injured two others
   because he thought the operation was a failure but no hospitals and doctors
   admitted that. After this case a large number of doctors and nurses in different
   cities across China protested in different ways, such as going into the streets,
   strikes and online protests.
8. Land acquisition and urban demolition has been one of the tensest issues over
   recent years. Some house owners, such as Tang Fuzhen in 2009, Xi Xinzhu
   in 2009, Tao Huixi in 2010, the family with the surname of Zhong in 2010,
   Wang Jiazheng in 2011, Shi Ganming in 2012, Zheng Guocun in 2013 and
   Feng Wenji and his wife in 2014, burned themselves as a protest against local
   authorities' ignoring their requests to talk and the violent demolition of their
   houses. In some cases, such as several murders in 2013, the Pingdu Tragedy in
   2014, land or house owners were even killed or severely injured by fires set by
   or bulldozers used by those carrying out the demolition.
9. The conflict between vendors and urban administrators has been fierce over
   recent years. A result of this is the violent killing of vendors by urban administrators or urban administrators by vendors. The former is exemplified in the
   case of Deng Zhengjia, who was a watermelon peasant and vendor and was
   killed by four urban administrators in 2013; while an example of the latter
   is the case of Xia Junfeng, who killed two urban administrators in 2009. The
   death of Deng Zhengjia resulted in a riot in the locality. The sentences of

the four killers were regarded as too light by online commentators and the commercial news media. Though Xiao Junfeng was sentenced to death and executed, sympathy was expressed towards him in the commercial media and online.

10. In April 2014, several thousands of workers in Dongguan City, Guangdong Province went on strike because of payment and pension disputes. The strike went on for about two weeks. Through national Party organs, the central government put pressure on local governments, urging them to handle the strikes and the wrongdoing of manufacturing enterprises properly and promptly.

11. "China's future: enter the Chinese NGO", in *The Economist*, April 12–18, 2014, 12–14.

1 Modernisation, Environmental Problems and Chinese Society

1. Accessed on August 23, 2012 at http://www.worldbank.org/en/country/china/overview.
2. From "How strong is China's economy?" accessed on August 23, 2012 at http://www.economist.com/node/21555915.
3. http://databank.worldbank.org, accessed on July 12, 2012.
4. "Energy for China" accessed on October 28, 2013, at http://www.economist.com/node/9488954.
5. "China may see worst energy crisis in years", accessed on October 28, 2013 at http://www.china.org.cn/china/2011–05/17/content_22577132.htm.
6. According to an article published by *Liaowang Oriental Weekly*: Mining areas in Baiying Gansu suffered from ground collapse: villages forced to leave, accessed on September 5, 2012 at http://news.sina.cn/c/sd/2009–07–20/141718258136.shtml.
7. According to an article published by *People's Daily*: An investigation into chemical projects and plants: severe environmental risks exist in most water areas, accessed on September 5, 2012 at http://news3.xinhuanet.com/newscenter/2006–07/12/content_4820023.htm.
8. According to the Bulletin of First National Census for Water, published by China's National Bureau of Statistics and the Ministry of Water Resources on March 2013. The data in the census were collected in 2011.
9. On May 18, 2011 the State Council's Committee Meeting chaired by the then Premier Wen Jiabao acknowledged the environmental problems and issues the Three Gorges dam has caused (Bristow 2011).
10. Many news media articles have reported on cancer villages in China and China's silence-breaking on this issue. See, for example, "China Releases Grim Cancer Statistics" http://mobile.businessweek.com/articles/2013–04–09/grim-cancer-statistics-from-china; "China's Top 6 Environmental Concerns" http://www.livescience.com/27862-china-environmental-problems.html, accessed on April 16, 2013; "Illegal toxic dump in China is a scandal too foul to be covered up", *The Times*, February 27, 2013.
11. Accessed on June 4, 2014 at http://www.caep.org.cn/uploadfile/greenGDP/gongzhongban2004.pdf

12. Accessed on September 5, 2012 at http://www.gov.cn/gzdt/2006-09/07/content_381190.htm.
13. http://jcz.cq.gov.cn/Html/1/czdt/gncj/2013-01-17/48898.html.
14. Accessed on September 6, 2012 at http://www.zhb.gov.cn/ztbd/gzhy/diqici-hbdh/ljhbdh/201112/t20111221_221578.htm.
15. Accessed on September 6, 2012 at http://www.zhb.gov.cn/ztbd/gzhy/diqici-hbdh/ljhbdh/201112/t20111221_221579.htm.
16. "Li Keqiang Commented on Environmental Protection: Government should deal strictly with environmental offences", accessed on October 30, 2013 at http://news.xinhuanet.com/politics/2013-03/17/c_115054263.htm.
17. http://www.gov.cn/gzdt/2013-04/19/content_2382198.htm.
18. http://paper.people.com.cn/rmrbhwb/html/2013-04/13/content_1225057.htm.
19. "Served in China", *The Economist*, February 23, 2013.
20. http://finance.people.com.cn/n/2013/0520/c348883-21539459.html.
21. http://finance.china.com.cn/news/gnjj/20130321/1343448.shtml.
22. Accessed on September 7, 2012 at http://www.gov.cn/zwgk/2009-05/18/content_1317790.htm.
23. Accessed on September 7, 2012 at https://sites.google.com/site/noguoguang/.
24. An article entitled "How the public opinion in Taiwan made 'Guoguang Petrochemical' abandoned", published by the *Southern Weekly*, Guangzhou, on May 23, 2011. Accessed on September 7, 2012 at http://www.infzm.com/content/59423.
25. An article named Oyster VS. Petrochemical Industry (*shenghao pk shihua*) published by *Xinmin Weekly* Magazine, on September 19th, 2011 accessed on September 7, 2012 at http://blog.sina.com.cn/s/blog_61fe8a7f0100y92t.html
26. http://kjs.mep.gov.cn/zghjbz/rzhcx/, accessed on April 8, 2013.
27. "Several advanced environmental protection technologies in Shijiazhuang Medicine Manufacturing Group are the first of this kind in China", published by *Shijiazhuang Daily* on the website of the Xinhuanet, see http://www.he.xinhuanet.com/news/2012-09/16/c_113093078.htm, accessed on April 8, 2013.
28. It doesn't matter what colour a cat is, so long as it catches mice.
29. http://www.ft.com/cms/s/0/aa8e771a-a29a-11e2-bd45-00144feabdc0.html#axzz2QdOXEPb1.
30. Accessed on September 10, 2012 at http://www.guardian.co.uk/environment/2011/apr/14/toxic-mine-spill-chinese-pollution.
31. Accessed on September 10, 2012 at http://www.guardian.co.uk/environment/2011/apr/14/toxic-mine-spill-chinese-pollution
32. http://www.chinadevelopmentbrief.org.cn/.
33. According to interviews with Qian Qiang and Mang Yong in 2011. All names of interviewees used in the book are pseudonyms.
34. "Blogging battle over Beijing smog", *The Guardian*, December 7, 2011.
35. A real estate developer.
36. "China warns US Embassy to stop reporting Beijing pollution", *The Independent*, June 5, 2012.

37. Accessed on September 12, 2012 at http://www.npc.gov.cn/npc/bmzz/ huanjing/2008–01/07/content_1388386.htm.
38. Accessed on September 12, 2012 at http://www.npc.gov.cn/npc/bmzz/ huanjing/2008–01/07/content_1388386.htm.
39. Accessed on September 12, 2012 at http://zls.mep.gov.cn/hjtj/ nb/2010tjnb/201201/t20120118_222718.htm.

2   Twenty Years of Environmental Investigative Reporting: Agendas, Social Interests and Voices

1. Five-Year Plan refers to a central planning of national economic development over a period of five years. The first "Five-Year Plan" was made and
   announced in 1955 for 1953–1957. To speed up industrialisation and to over-
   take the United States and the United Kingdom were the aim of the first Five-
   Year Plan, while the Great Leap Forward was the goal of the second Five-Year
   Plan. The most recent (the twelfth) plan was made for 2011–2013 in 2010
   with the aim of deepening economic reforms and accelerating the transformation of economic development.
2. Data was collected from the Wisers database on April 14, 2014.
3. In the case of Tibetan antelope (chiru), NGOs, especially Friend of Nature,
   played a prominent role. Tibetan antelope is a species of antelope that lives
   in Hoh Xil (the Qinghai–Tibetan plateau) and is a national protected species.
   These animals have long and beautiful fur. In the 1990s the number of Tibetan
   antelopes dropped rapidly due to the shahtoosh trade (they were extensively
   and illegally hunted (overhunting) for their hair (shahtoosh) and skins to
   meet the demand in the Western market (*The Economist* 2010). From 1992
   to 1999, the Qinghai government caught 114 illegal hunters of Tibetan antelope and this involved 15,243 pieces of Tibetan antelope skins (Fang and Ye
   2000). It has been estimated and recorded that some 20,000 Tibetan antelopes
   have been killed annually for manufacturing commercial shahtoosh. From
   1996, Friend of Nature appealed to stop the illegal hunting and trade and
   to save Tibetan antelopes under threat (they wrote to Tony Blair, the then
   prime minister of the United Kingdom and submitted a report to the state
   Environmental Protection Bureau and the state Forest Bureau). Several years
   later, news media such as CCTV's *Oriental Horizon* (*dongfang shikong*), *Focus*
   and *Southern Weekend* started to give attention to the issue in the late 1990s
   and at the turn of the new century. Xi Zhinong, who initiated the protection
   of the snub-nosed monkey, joined *Oriental Horizon* at CCTV and investigated
   the situation of Tibetan antelopes in 1997 and 1999. He reported on this
   issue through a documentary. In 2000, *Southern Weekend* published a famous
   investigative report, "Who will protect the Hoh Xil?", which was adapted
   into an award-winning film, *Hoh Xil*, by Lu Chuan. It was seen as an NGO
   and media success.The campaign was somewhat effective, as the state started
   taking official action toward the illegal hunting and trade. In 1999, the state
   Forest Bureau and the police in the regions of Qinghai, Xinjiang and Tibet
   launched the "Hoh Xil" No 1 Action to crackdown on the illegal hunting
   of Tibetan Antelopes, which were defined as criminal activities. However,
   the success may be credited to the resistance on a global scale/global ban on

Shahtoosh trade as well as the fact that the hunting and trade itself is illegal. As early as 1975, under the Convention on International Trade in Endangered Species (CITES), the Shahtoosh trade was banned globally. Internationally, for example, The World Wildlife Fund for Nature and Traffic India launched the Say No to Shahtoosh campaign in 1999. http://www.indiaenvironmentportal. org.in/content/103757/campaign-to-ban-shahtoosh-shawl-launched/

4. According to the local *Yearbook of Bingzhou City*, Shandong Province, accessed on November 7, 2013 at the government website: http://sd.infobase.gov.cn/ bin/mse.exe?seachword=&K=bd&A=3&rec=80&run=13

5. Mu is a Chinese measurement unit, which is equal to 666.5 square meters.

6. The "Restoring Farmland to Forest" project was started in 1999 in several cities such as Yan'an and nationwide in 2002. The current Regulation of Restoring Farmland to Forests was issued in 2002 and enforced in 2003. Accessed on November 11, 2013, at http://www.gov.cn/wszb/zhibo515/content_2152520. htm and http://news.xinhuanet.com/fortune/201308/25/c_117079497.htm.

7. Accessed on November 11, 2013 at http://tghl.forestry.gov.cn/portal/tghl/ s/2423/content-525274.html.

8. This figure was cited in a report titled "Smog saved air-clean equipment market", by Bingqun Duan and Zhong'an Zhang, at the *Guangzhou Daily*, January 8, 2014.

9. Accessed on November 11, 2013, at http://www.gov.cn/zwgk/2007–12/24/ content_841978.htm.

10. In 2009 and 2010, a series of in-depth reports by Tang Yaoguo published at *Liaowang* Magazine for which he worked outlined the history of the debates surrounding the construction of the Three Gorges Dam.

11. http://www.xys.org/dajia/nujiang.html.

3 The Discourse of Risk: Environmental Problems and Environmentalism in Chinese Press Investigative Reports

1. Ten newspapers have been selected for this study. They are selected because they have a reputation for practicing investigative journalism and an enduring interest in the topic of environmental problems and issues. The ten newspapers are *Southern Weekend*, *Southern Metropolitan Daily*, *Yunnan Information*, *Dahe Daily*, *Xiaoxiang Morning*, *Oriental Morning*, *Beijing Youth*, *Beijing News*, *First Economic Daily* and *People's Daily*. Their background can be found in Table 3.1. These outlets are based on different geographical locations and are different in terms of editorial principles and readership. The articles were collected through the Wisers database which is a Chinese news database based in Hong Kong. In total, the sample corpus includes 258 investigative reports. Part of the empirical findings in this chapter, in particular some findings of framing analysis, has been published in "Environmental Risks in Newspaper Coverage: A Framing Analysis of Investigative Reports on Environmental Problems in 10 Chinese Newspapers", in *Environmental Communication*, 8(3), 345–367.

2. Discourse analysis in this chapter adopts the analytical approach suggested by Fairclough (Fairclough 2000, 1995, 1989), especially examining the appearance of meaningful key lexical words in investigative reports.

3. For framing analysis, this chapter follows and develops from the under-standing of paradigm and functions of framing initiated by Entman (Entman 1993) and the categorisation and measurement of framing developed by Semetko and Valkenburg (Semetko and Valkenburg 2000) and used by Dirikx and Gelders (Dirikx and Gelders 2010).
4. Pishuang is Chinese popular name for Shen, which is the official name.
5. Here the coverage of nationwide press means the coverage of hard news.

4   Environmental Investigative Journalists and Their Work

1. The name of the news outlet was removed in order to keep the investigative journalist anonymous.
2. The name of the news outlet was removed in order to keep the investigative journalist anonymous.

5   Offline Investigative Journalism and Online Environmental Crusades

1. http://news.china.com.cn/live/2013–02/19/content_18701613.htm.
2. http://www.infzm.com/content/98452.
3. According to an article entitled "Shandong Environmental Protection Department: Deng Fei Should Apologise As Enterprises Wronged of Discharging Waste Water Underground" (September 2, 2013) published by *Southern Metropolitan Daily*, accessed at http://wen.oeeee.com/a/20130902/1072454.html.
4. According to an article with the title of "Environmental Protection Department Director Reflecting on the Event of Wastewater Discharging Through High-Pressure Wells" (February 28, 2014), accessed at http://www.infzm.com/content/98452.
5. http://www.mep.gov.cn/gkml/hbb/qt/201305/t20130509_251858.htm.
6. http://news.xinhuanet.com/politics/2013–09/14/c_125386974.htm; http://sd.sina.com.cn/news/sdyw/2013–12–06/074950569.html
7. According to an article titled "Big Vs Near Dusk" (September 12, 2013). Accessed at http://www.infzm.com/content/94222.
8. http://www.infzm.com/content/94222.
9. http://online.wsj.com/news/articles/SB10001424127887324178904578341433004369650.
10. http://www.topnews9.com/spec/20130219/17214.html.
11. http://view.news.qq.com/zt2013/wfswr/index.htm.
12. http://www.cnnic.cn/hlwfzyj/hlwxzbg/hlwtjbg/201301/P020130122600399530412.pdf.
13. http://ep.chinaluxus.com/Efs/20120531/186583.html.
14. http://news.sciencenet.cn/htmlnews/2012/4/263244.shtm.
15. Qiqi's Heaven is a non-government blog set up for memorialising the extinct species of baiji and for saving the species of porpoise that is near extinction. This blog has become an important part of porpoise protection. More details to be found on the website: http://blog.sina.com.cn/u/1491304320.

16. http://news.xinhuanet.com/2012–04/18/c_111799348.htm.
17. http://news.sina.com.cn/c/2013–06–29/023927527627.shtml.
18. http://news.sina.com.cn/o/2013–02–25/045926347733.shtml.
19. http://zhongwaiduihua.blog.caixin.com/archives/58274.

6 Hegemony and Counter-Hegemony: Investigative Journalism between Modernisation and Environmental Problems

1. http://news.xinhuanet.com/ziliao/2003–05/30/content_894678.htm.
2. http://news.xinhuanet.com/ziliao/2003–05/30/content_894678.htm.

Bibliography

Althusser, L. (1971). *Lenin and Philosophy, and Other Essays*. New York: Monthly Review Press; London: New Left Books.

Beck, U. (1992). *Risk Society: Towards a New Modernity*. London, New Delhi, Thousand Oaks: SAGE Publications.

—— (1996). "Risk Society and the Provident State". In *Risk, Environment, and Modernity: Towards a New Ecology*. S. Lask, B. Szerszynski and B. Wynne. London, Thousand Oaks and New Delhi: SAGE Publications, 27–43.

Beck, U., A. Giddens, et al. (1994). *Reflexive Modernization: Politics, Tradition, and Aesthetics in the Modern Social Order*. Stanford: Stanford University Press.

Béja, J.-P. (2006). "The Changing Aspects of Civil Society in China". *Social Research* 73(1): 53–74.

Benton, T. (1996). "Introduction to Part 1". In *The Greening of Marxism*. T. Benton. New York: The Guilford Press, 7–16.

Berger, P. and T. Luckmann (1967). *The Social Construction of Reality*. New York: Anchor Books.

Bo, Z. (2008). "Balance of Factional Power in China: The Seventeenth Central Committee of the Chinese Communist Party". *East Asia* 25(4): 333–364.

Boucher, D. (1996). "Not with a Bang but a Whimper". *Science and Society* 60(3): 279–289.

Brand, P., M. J. Thomas, et al. (eds) (2005). *Urban Environmentalism: Global Change and the Mediation of Local Conflict*. London: Routledge.

Bristow, M. (2011). "China Acknowledges Three Gorges Dam 'Problems'". BBC. 19 May 2011. http://www.bbc.co.uk/news/world-asia-pacific-13451528

Cai, Y. (2014). "Managing Group Interests in China". *Political Science Quarterly* 129(1): 107–131.

Cao, P. (2009). "Financial Policies and Measures Expected to Save the Press in Winter (*baoye guodong youdai chutai jiushi cuoshi*)". *Journalists* (*xinwen jizhe*) 1(311): 28–30.

Cao, Q. (2001). "Journalism as Politics: Reporting Hong Kong's Handover in the Chinese Press". *Journalism Studies* 1(4): 665–678.

Castells, M. (1996). *The Rise of The Network Society: The Information Age: Economy, Society and Culture*. Cambridge, MA, and Oxford, Blackwell Publishers.

Chin, J. and B. Spegele (2013). "China's Bad Earth". *The Wall Street Journal*. Washington, DC. 27 July 2013. http://online.wsj.com/articles/SB10001424127887323382910457862401064822 8142

Christoff, P. (1996). "Ecological Modernization, Ecological Modernities". *Environmental Politics* 5(3): 476–500.

Chu, H. (1998). "China Dam Damned: Are Disastrous Floods in China Being Caused by a Huge Dam Building Project?" *The Guardian*. London. 5 August 1998.

CNNIC (2014). *33rd Statistical Report on the State of the Chinese Internet's Development*. CNNIC. Beijing.

Cohen, S. and J. Young, (eds) (1973). *The Manufacture of News: Social Problems, Deviance and the Mass Media*. London, Constable.

Cottle, S. (2000). "TV News, Lay Voices and the Visualisation of Environmental Risks". In *Environmental Risks and the Media*. A. Stuart, A. Barbara and C. Cynthia. London and New York, Routledge: 29–44.

—— (2006). *Mediatized Conflict*. Maidenhead, Open University Press.

De Burgh, H. and R. Zeng (2012). "Environment Correspondents in China in Their Own Words: Their Perceptions of Their Role and the Possible Consequences of Their Journalism". *Journalism* 13(8): 1004–1023.

Dirikx, A. and D. Gelders (2010). "To Frame is to Explain: A Deductive Frame-Analysis of Dutch and French Climate Change Coverage during the Annual UN Conferences of the Parties". *Public Understanding of Science* 19: 732–742.

Dong, S. and J. Yang (2007). "Modernity, Utopia, and the Development of Socialism in China (*xiandai xing, wutuobang, zhongguo shehui zhuyi fazhan licheng*)". *Journal of Henan University: Social Science (henan daxue xuebao shehui kexueban)* 47(6): 86–89.

Douglas, M. and A. Wildavsky (1982). *Risk and Culture: An Essay on the Selections of Technological and Environmental Dangers*. Berkeley, University of California Press.

Doyle, T. and D. McEachern (2007). *Environment and Politics*. London, Routledge.

Dryzek, J. S. (2005). *The Politics of the Earth: Environmental Discourses*. Oxford and New York, Oxford University Press.

Dudgeon, D. (1995). "River Regulation in Southern China: Ecological Implications, Conservation and Environmental Management". *Regulated Rivers: Research & Management* 11(1): 35–54.

Economy, E. C. (2006). "Environmental Governance: The Emerging Economic Dimension". *Environmental Politics* 15(2): 171–189.

—— (2010). *The River Runs Black: The Environmental Challenge to China's Future*. New York, Cornell University Press.

Eder, K. (1996). "The Institutionalisation of Environmentalism: Ecological Discourse and the Second Transformation of the Public Sphere". *Risk, Environment and Modernity: Towards a New Ecology*. S. Lash, B. Szerszynski and B. Wynne. London, Thousand Oaks and New Delhi, SAGE Publications: 203–223.

Entman, R. M. (1993). "Framing: Toward Clarification of a Fractured Paradigm". *Journal of Communication* 43: 51–58.

Ettema, J. and T. Glasser (1998). *Custodians of Conscience: Investigative Journalism and Public Virtue*. New York, Columbia University Press.

Fairclough, N. (1989). *Language and Power*. London, Longman.

—— (1992). *Discourse and Social Change*. Cambridge, Oxford, and Malden, Polity Press.

—— (1995). *Media Discourse*. London, Edward Arnold.

—— (2000). *New Labour, New Language*. London and New York, Routledge.

Fan, Y. (2013). "My Views on the Merger of Two Press Groups in Shanghai (*shanghai liangda baoye hebing zhi wojian*)". *Southern Media Studies (nanfang chuanmei yanjiu)* 45: 28–33.

Fang, S. and H. Ye (2000). "Who Protects Hoh Xil?" *Southern Weekend*.

Foster, J. B. (1999a). "Marx's Theory of Metabolic Rift: Classical Foundations for Environmental Sociology". *American Journal of Sociology* 105(2): 366–405.
—— (1999b). *The Vulnerable Planet: A Short Economic History of the Environment*. New York, Monthly Review Press.
—— (2000). *Marx's Ecology: Materialism and Nature*. New York, Monthly Review Press.
—— (2010). "Why Ecological Revolution?" *Monthly Review* 61(8).
Foster, J. B., B. Clark, et al. (2010). *The Ecological Rift: Capitalism's War on the Earth*. New York, Monthly Review Press.
Foucault, M. (1970). *The Order of Things: An Archaeology of the Human Sciences*. London, Tavistock.
—— (1972). *The Archaeology of Knowledge*. London, Routledge.
—— (1979). *Discipline and Punish: The Birth of the Prison*. New York, Vintage.
Fowler, R. (1991). *Language in the News: Discourse and Ideology in the Press*. London, Routledge.
Fu, B.-J., X.-L. Zhuang, et al. (2007). "Environmental Problems and Challenges in China". *Environmental Science & Technology* 41: 7597–7602.
Galtung, J. and M. H. Ruge (1965). "The Structure of Foreign News: The Presentation of the Congo, Cuba and Cyprus Crises in Four Norwegian Newspapers". *Journal of International Peace Research* 2(1): 64–90.
Gare, A. (1995). *Postmodernism and the Environmental Crisis*. London, Routledge.
Giddens, A. (1981). *A Contemporary Critique of Historical Materialism. Volume 1: Power, Property and the State*. Houndmills, Basingstoke and London, Macmillan Education Ltd.
—— (1990). *The Consequences of Modernity*. Cambridge, Polity Press.
—— (1991). *Modernity and Self-Identity: Self and Society in the Late Modern Age*. Cambridge, Polity.
Goldman, P. (2006/2007). "Public Interest Environmental Litigation in China: Lessons Learned from the U.S. Experience". *Vermont Journal of Environmental Law* 8: 251–280.
Gramsci, A. (1971). *Prison Notebooks*. New York, International Publishers.
—— (1977). *Selections from Political Writings, 1910–1920*. London, GBR Electric Book Company.
Guo, P. (2005). *Corporate Environmental Reporting and Disclosure in China*. Ed. R. Welford. Beijing, Tsinghua University, CSR Asia.
Guo, S. and J. J. Kassiola, (eds) (2010). *China's Environmental Crisis: Domestic and Global Political Impacts and Responses*. London and New York, Palgrave Macmillan.
Habermas, J. (1989). *The Structural Transformation of the Public Sphere*. Cambridge, Polity Press.
Hajer, M. (1995). *The Politics of Environmental Discourse: Ecological Modernisation and the Policy Process*. New York and London, Oxford University Press.
Hall, S. (1992). "The West and the Rest". In *Formations of Modernity*. B. Gieben and S. Hall. Cambridge, Polity Press/The Open University.
Hannigan, J. (2006). *Environmental Sociology*. London and New York, Routledge.
Hansen, A. (1991). "The Media and the Social Construction of the Environment". *Media Culture and Society* 13: 443–458.
—— (2010). *Environment, Media and Communication*. London, Routledge.

Harcup, T. and D. O'Neill (2001). "What Is News? Galtung and Ruge Revisited". *Journalism Studies* 2(2): 261–280.

Harvey, D. (1990). *The Condition of Postmodernity: An Enquiry into the Origins of Cultural Change.* Oxford, Blackwell.

He, F. (2009). "Price of Prosperity: Economic Development and Biological Conservation in China". *Journal of Applied Ecology* 46: 511–515.

Ho, P. (2001). "Greening without Conflict? Environmentalism, NGOs and Civil Society in China". *Development and Change* 32: 893–921.

—— (2006). "Trajectories for Greening in China: Theory and Practice". *Development and Change* 37(1): 3–28.

—— (2008). "Introduction". In *China's Embedded Activism: Opportunities and Constraints of a Social Movement.* P. Ho and R. L. Edmonds. London and New York, Routledge Studies on China in Transition: 1–19.

Ho, P. and R. L. Edmonds, (eds) (2008). *China's Embedded Activism: Opportunities and Constraints of a Social Movement.* London and New York, Routledge Studies on China in Transition.

Ho, W.-C. (2012). "The Rise of the Bureaucratic Bourgeoisie and Factional Politics of China". *Journal of Contemporary Asia* 42(3): 514–521.

—— (2013). "What Analyses of Factional Politics of China Might Miss When the Market Becomes a Political Battlefield: The Telecommunication Sector as a Case in Point". *China Review* 13(1): 71–92.

Hodgson, V. and M. Foley, (eds) (2003). *The Civil Society Reader.* Hanover, NH, Tufts University Press.

Holbig, H. and B. Gilley (2010). "Reclaiming Legitimacy in China". *Politics and Policy* 38(3): 395–422.

Horowitza, S. and C. Marshb (2002). "Explaining Regional Economic Policies in China: Interest Groups, Institutions, and Identities". *Communist and Post-Communist Studies* 35(2): 115–132.

Howell, J. (2012). "Civil Society, Corporatism and Capitalism in China". *Journal of Comparative Asian Development* 11(2): 271–297.

Huan, Q. (2007). "Ecological Modernisation: A Realistic Green Road for China?" *Environmental Politics* 16(4): 683–687.

Jahiel, A. R. (2006). "China, the WTO, and Implications for the Environment". *Environmental Politics* 15(2): 310–329.

Jing, J. (2010). "Environmental Protests in Rural China". In *Chinese Society: Change, Conflict and Resistance.* E. J. Perry and M. Selden. Oxon and New York, Routledge: 197–214.

Jing, Y. (2013). "The One-Child Policy Needs an Overhaul". *Journal of Policy Analysis and Management* 32(2): 392–399.

Kaviraj, S. and S. Khilnani (2002). *Civil Society: History and Possibilities.* Cambridge, Cambridge University Press.

Layfield, D. (2008). *Marxism and Environmental Crises.* Bury St. Edmunds, Arena Books.

Lee, C.-C., Z. He, et al. (2006). "Chinese Party Publicity Inc.: Conglomerated: The Case of the Shenzhen Press Group". *Media, Culture & Society* 28(4): 581–602.

—— (2007). "Party-Market Corporatism, Clientelism, and Media in Shanghai". *The International Journal of Press/Politics* 12(3): 21–42.

Leibold, J. (2011). "Blogging Alone: China, the Internet, and the Democratic Illusion?" *The Journal of Asian Studies* 70: 1023–1041.

Lester, L. (2010). *Media and Environment*. Cambridge and Malden, Polity Press.

Li, C. (2009). "China's Team of Rivals". *Foreign Policy* **171**: 88–93.

Li, W. and D. T. Yang (2005). "The Great Leap Forward: Anatomy of a Central Planning Disaster". *Journal of Political Economy* **113**(4): 840–877.

Li, X. (2002). "'Focus' (Jiaodian Fangtan) and the Changes in the Chinese Television Industry". *Journal of Contemporary China* **11**(30): 17–34.

Lia, V. and G. Langa (2010). "China's 'Green GDP' Experiment and the Struggle for Ecological Modernisation". *Journal of Contemporary Asia* **40**(1): 44–62.

Liang, H. (2012). "Mine Inspection Right vs. 80 Snub-Nosed Monkeys: What Would You Choose (*tankuang quan vs. 80 dianjinsihou ni hui zuo he xuanze*)?" *Yunnan Information*, Yunnan Information. 19 June 2012.

Liu, C. (1998). "Environmental Issues and the South-North Water Transfer Scheme". *The China Quarterly* **156**: 899–910.

Liu, C., J. Yu, et al. (2001). "Groundwater Exploitation and Its Impact on the Environment in the North China Plain". *Water International* **26**(2): 265–272.

Liu, C. and H. Zheng (2002). "South-to-North Water Transfer Schemes for China". *International Journal of Water Resources Development* **18**(3): 453–471.

Liu, J. (2007). "The 'special interests' destroying China's environment". *Chinadialogue*. Beijing. 24 January 2007. https://www.chinadialogue.net/article/715-The-special-interests-destroying-China-s-environment

Liu, J. and J. Diamond (2008). "Revolutionizing China's Environmental Protection". *Science* **319**: 37–38.

Liu, L. (2010). "Made in China: Cancer Villages". *Environment: Science and Policy for Sustainable Development* **52**(2): 8–21.

Lo, C. W. H. and S. W. Leung (2000). "Environmental Agency and Public Opinion in Guangzhou: The Limits of a Popular Approach to Environmental Governance". *The China Quarterly* **163**: 677–704.

Lo, C. W. H., P. K. T. Yip, et al. (2000). "The Regulatory Style of Environmental Governance in China: The Case of EIA Regulation in Shanghai". *Public Administration and Development* **20**: 294–207.

Lu, B. (2009). "Interview Water Crisis: Digging the Rich Mine of the Truth (*caifang shui weiji kaicai zhenxiang de fukuang*)". *Young Journalists* (*qingnian jizhe*) **7**: 42–43.

Lu, C. (2013). "China's Middle Class: Unified or Fragmented?" *Japanese Journal of Political Science* **14**(1): 127–150.

Lu, Y. (2007). "Environmental Civil Society and Governance in China". *International Journal of Environmental Studies* **64**(1): 59–69.

Lum, T. (2006). "Internet Development and Information Control in the People's Republic of China", Washington, DC, Congressional Research Service, The Library of Congress.

Ma, C. (2010). "Who Bears the Environmental Burden in China – An Analysis of the Distribution of Industrial Pollution Sources?" *Ecological Economics* **69**(9): 1869–1876.

Ma, X. and L. Ortolano (2000). *Environmental Regulation in China: Institutions, Enforcement, and Compliance*. Oxford, Rowman & Littlefield.

McAllister, D. E., J. F. Craig, et al. (2001). "Biodiversity Impacts of Large Dams", Background Paper Number 1. Prepared for IUCN/UNEP/WCD.

MacBean, A. (2007). "China's Environment: Problems and Policies". *The World Economy* 30(2): 292–307.

McNair, B. (2003). *Sociology of Journalism*. London, Routledge.

—— (2006). *Cultural Chaos: Journalism, News and Power in a Globalised World*. London, Routledge.

Marqu, C., J. Zhang, et al. (2011). "Regulatory Uncertainty and Corporate Responses to Environmental Protection in China". *California Management Review* 54(1): 39–63.

Martens, S. (2006). "Public Participation with Chinese Characteristics: Citizen Consumers in China's Environmental Management". *Environmental Politics* 15(2): 211–230.

Marx, K. (1976). *Capital*. London, Penguin Books.

Meyers, O. (2007). "Memory in Journalism and the Memory of Journalism: Israeli Journalists and the Constructed Legacy of Haolam Hazeh". *Journal of Communication* 57(4): 719–738.

Miles, L. and R. Croucher (2013). "Gramsci, Counter-hegemony and Labour Union: Civil Society Organisation Coalitions in Malaysia". *Journal of Contemporary Asia* 43(3): 413–427.

Mol, A. P. J. (2006). "Environment and Modernity in Transitional China: Frontiers of Ecological Modernization". *Development and Change* 37(1): 29–56.

—— (2009). "Environmental Governance through Information: China and Vietnam". *Singapore Journal of Tropical Geography* 30: 114–129.

Molotch, H. and M. Lester (1975). "Accidental News: The Great Oil Spill as Local Occurrence and National Event". *American Journal of Sociology* 81(2): 235–260.

Morris, J. T. (2012). *Risk, Language, and Power: The Nanotechnology Environmental Policy Case*. Plymouth, Lexington Books.

Muldavin, J. (1998). "Agrarian Change in Contemporary Rural China. In *Privatising the Land: Rural Political Economy in Post-Communist Societies*. I. Szelenyi. London and New York, Routledge: 92–124.

Nayar, P. K. (2010). *Contemporary Literary and Cultural Theory: From Structuralism to Ecocriticism*. New Delhi, Dorling Kindersley Pvt. Ltd.

Neuzil, M. (2008). *The Environment and The Press: From Adventure Writing to Advocacy*. Evanston, IL, Northwestern University Press.

Ngai, P. (2008). "Subsumption or Consumption? The Phantom of Consumer Revolution in 'Globalizing' China". *Cultural Anthropology* 18(4): 469–492.

Nickum, J. E. and Y.-S. F. Lee (2006). "Same Longitude, Different Latitudes: Institutional Change in Urban Water in China, North and South". *Environmental Politics* 15(2): 231–247.

O'Shannassy, M. (2009). "Beyond the Barisan Nasional?: A Gramscian Perspective of the 2008 Malaysian General Election". *Contemporary Southeast Asia: A Journal of International and Strategic Affairs* 31(1): 88–109.

Oakes, T. (2004). "Building a Southern Dynamo: Guizhou and State Power". *The China Quarterly* 178: 467–487.

Oi, J. C. (1985). "Communism and Clientelism: Rural Politics in China". *World Politics* 37(2): 238–266.

Palmlund, I. (1992). "Social Drama and Risk Evaluation". In *Social Theories of Risk*. S. Krimsky and D. Golding. Westport, CT, Praeger.

Perelman, M. (1996). "Marx and Resource Scarcity". In *The Greening of Marxism*. T. Benton. New York, The Guilford Press: 60–80

Qie, J. (2007). "Witnessing the Up and Down of Environmental Reporting (*jianzheng huanbao xinwen de leng yu re*)". *Chinese Journalist* (*zhongguo jizhe*) **4**: 40–41.

Qin, B., P. Xu, et al. (2007). "Environmental Issues of Lake Taihu, China". *Developments in Hydrobiology* **194**(2): 3–14.

Qiu, J. L. (2009). *Working-class Network Society: Communication Technology and the Information Have-Less in Urban China*. Cambridge, MA and London, The MIT Press.

Qu, G. (2010). *China's Environmental Protection Strategies* (*zhongguo huanjing baohu fanglue*). Beijing, China Environmental Science Publisher (*zhongguo huanjing kexue chubanshe*).

Rozman, G. (ed.) (1981). *The Modernisation of China*. New York, Free Press.

Sanders, R. (1999). "The Political Economy of Chinese Environmental Protection: Lessons of the Mao and Deng Years". *Third World Quarterly* **20**(6): 1201–1214.

Schnaiberg, A. (1980). *The Environment: From Surplus to Scarcity*. New York, Oxford University Press.

Schudson, M. (2003). *The Sociology of News*. New York, W. W. Norton & Company.

Schwartz, B. (1965). "Modernisation and the Maoist Vision – Some Reflections on Chinese Communist Goals". *The China Quarterly* **21**: 3–19.

Schwartz, D. (2006). *Writing Green: Advocacy and Investigative Reporting about the Environment in the Early 21st Century*. Baltimore, Apprentice House.

Semetko, H. A. and P. M. Valkenburg (2000). "Framing European Politics: A Content Analysis of Press and Television News". *Journal of Communication* **50**(2): 93–109.

Shapiro, J. (2001). *Mao's War Against Nature*. Cambridge, Cambridge University Press.

—— (2012). *China's Environmental Challenges*. Cambridge and Malden, Polity Press.

Shen, J., Y. Pan, et al. (2009). "Impacts of Hekou Hydropower Station on Fish Resources and Suggestions for Fish Protection (*hekou shuidianzhan dui yulei ziyuan de yingxiang ji baohu cuoshi*)". *People's Yellow River* (*renmin huanghe*) **31**(8): 51–52.

Shi, H. and L. Zhang (2006). "China's Environmental Governance of Rapid Industrialisation". *Environmental Politics* **15**(2): 271–292.

Simon, K. W. (2013). *Civil Society in China: The Legal Framework from Ancient Times to the "New Reform Era"*. Oxford and New York, Oxford University Press.

Smil, V. (1996). "Environmental Problems in China: Estimates of Economic Costs". East–West Center special reports; no. 5. Honolulu, East–West Center.

Sonnenfeld, D. A. and A. P. J. Mol (2002). "Globalization and the Transformation of Environmental Governance: An Introduction". *American Behavioral Scientist* **45**(9): 1318–1339.

Spires, A. J. (2012). "Lessons from Abroad: Foreign Influences on China's Emerging Civil Society". *The China Journal* **68**: 125–146.

Spyridou, L.-P., M. Matsiola, et al. (2013). "Journalism in a State of Flux: Journalists as Agents of Technology Innovation and Emerging News Practices". *International Communication Gazette* **75**(1): 76–98.

Steinberg, D. A. and V. C. Shih (2012). "Interest Group Influence in Authoritarian States: The Political Determinants of Chinese Exchange Rate Policy". *Comparative Political Studies* **45**(11): 1405–1434.

Stuart, A., A. Barbara, et al. (2000). "Introduction". *Environmental Risks and the Media*. A. Stuart, A. Barbara and C. Cynthia. London and New York, Routledge.

Sun, H., B. Gong, et al. (2013). "Ecological and Environmental Problems Caused by Hydropower Stations in Rural Areas and Discussions on Reimbursement (*nongcun shuidianzhan yinqi de shengtai huanjing wenti ji buchang cuoshi chutan*)". *Journal of Yangtze River Scientific Research Institute (changjiang kexue xueyuan xuebao)* 30(3): 12–15.

Sun, Y. and D. Zhao (2007). "Multifaceted State and Fragmented Society: The Dynamics of the Environmental Movement in China". *Discontented Miracle: Growth, Conflict and Institutional Adaptations in China*. D. L. Yang. River Edge, NJ, USA World Scientific: 111–160.

Svensson, M., E. Saether, et al., eds (2014). *Chinese Investigative Journalists' Dreams: Autonomy, Agency and Voice*. Lanham, Boulder, New York, Toronto and Plymouth, Lexington Books.

Tang, Y. (2009). "A Story of Debates on the Three-Gorge Dam (*sanxia lunzhan fengyun lu*)". *The Liaowang Magazine*, Beijing. 49: 15–19.

Tapsell, R. and J. Eidenfalk (2013). "Australian Reporting from East Timor 1975–1999: Journalists as Agents of Change". *Australian Journal of Politics and History* 59(4): 576–592.

The Ministry of Water Resources of the People's Republic of China (1999). *The 1998 Big Floods in China (zhongguo 98 da hongshui)*. Beijing, China Water Resources and Water-electricity Publisher.

Tilt, B. (2009). *Struggling for Sustainability in Rural China: Environmental Values and Civil Society*. New York, Columbia University Press.

Tong, J. (2010). "The Crisis of the Centralized Media Control Theory: How Local Power Controls Media in China". *Media Culture and Society* 32(6).

—— (2011). *Investigative Journalism in China: Journalism, Power, and Society*. New York and London, Continuum.

—— (2013). "The Importance of Place: An Analysis of Changes in Investigative Journalism in Two Chinese Provincial Newspapers". *Journalism Practice* 7(1): 1–16.

—— (2014). "Environmental Risks in Newspaper Coverage: A Framing Analysis of Investigative Reports on Environmental Problems in 10 Chinese Newspapers". *Environmental Communication*,8(3): 345–367.

Tong, J. and C. Sparks (2009). "Investigative Journalism in China Today". *Journalism Studies* 10(3).

Tse, D. K., R. W. Belk, et al. (1989). "Becoming a Consumer Society: A Longitudinal and Cross-Cultural Content Analysis of Print Ads from Hong Kong, the People's Republic of China, and Taiwan". *Journal of Consumer Research* 15(4): 457–472.

Tuchman, G. (1978). *Making News: A Study in the Construction of Reality*. New York, The Free Press.

Van Dijk, T. A. (1988). *News Analysis: Case Studies of International and National News in the Press*. Hillsdale, NJ, Hove, and London, Lea Lawrence Erlbaum Associates.

Varis, O. and P. Vakkilainen (2001). "China's 8 Challenges to Water Resources Management in the First Quarter of the 21st Century". *Geomorphology* 41: 93–104.

Waisbord, S. (2011). "Journalism, Risk, and Patriotism". In *Journalism After September 11*. B. Zelizer and A. Stuart. London and New York, Routledge: 273–291.

Walker, M. (1996). "China and the New Era of Resource Scarcity". *World Policy Journal* 13(1): 8–14.

Wang, A. (ed.) (2008). *What Supports the Mansion of Environmental Protection*. Beijing, Sanlian Publisher.

Wang, C. (2007). "Chinese Environmental Law Enforcement: Current Deficiencies and Suggested Reforms." *Vermont Journal of Environmental Law* 8(2): 159–194.

Wang, H. (2003). *China's New Order: Society, Politics, and Economy in Transition*. Cambridge, MA, London, the President and Fellows of Harvard College.

—— (2005). "China Today: Misguided Socialism Plus Crony Capitalism". *New Perspectives Quarterly* 22(2): 32–34.

—— (2006). "Enterprise Transformation and the Historical Fate of the Chinese Working Class: An Investigation Report into the Tranformation of Jiangsu Tongyu Group Company". *Frontier (Tianya)* 1.

Wang, H. and R. E. Karl (1998). "Contemporary Chinese Thought and the Question of Modernity". *Social Text: Intellectual Politics in Post-Tiananmen China* 55: 9–44.

Wang, Q.-J. (2005). "Transparency in the Grey Box of China's Environmental Governance: A Case Study of Print Media Coverage of an Environmental Controversy from the Pearl River Delta Region". *The Journal of Environment Development* 14: 278–312.

Wang, S. and S. Guo (2011). "Research on Government's Environmental Responsibility of the Resource-Exhausted City". *Journal of Political Science and Law (zhengfa luncong)* 5: 112–120.

Wang, X. (2010). "Dependent Rural Areas and Environmental Degradation (lunwen fuyong de nongcun yu huanjing ehua)". *Academia Bimestris (xuehai)* 2: 60–62.

Wang, Y. (2005). "Opposing Dam or Building Dam? Analysis on the International Anti-Dam Movement and Policy Recommendations for China *(jianba haishi fanba? guoji fanba yundong fansi yu woguo gonggong zhengce tiaozheng)*". *China Soft Science (zhongguo ruan kexue)* 8: 33–39.

Weart, S. (1988). *Nuclear Fear*. Massachusetts and London, Harvard University Press.

Wiest, N. C. (2001). "Green Voices in Greater China: Harmony and Dissonance". *Hong Kong Conference Report: Environmental Journalism in Mainland China, Taiwan and Hong Kong*. Hong Kong, Wilson Center.

Wu, F. (2003). "Environmental GONGO Autonomy: Unintended Consequences of State Strategies in China". *The Good Society* 12(1): 35–45.

Wyss, B. (2008). *Covering the Environment: How Journalists Work the Green Beat*. London, Routledge.

Xiang, L. (2012). "China and the 'Pivot'". *Survival: Global Politics and Strategy* 54(5): 113–128.

Xiao, G. (2002). "Neo-Leftists and Differentiation of Thoughts Among Intelligentsias in Contemporary China (xinzuopai yu dangdai zhongguo zhishi fenzi de sixiang fenhua)". *Contemporary China Studies (dangdai zhongguo yanjiu)* 76(1). http://www.modernchinastudies.org/cn/issues/past-issues/76-mcs-2002-issue-1/1219–2012–01–06–08–38–50.html

—— (2003). "Political Liberalists and Conservatives in Post-Reform China: The Collision of Polar Opposites and Its Historic Consequences (zhongguo gaige kaifang yilai zhengzhi zhong de ziyoupai yu baoshou pai liangji chongtu yiji lishi houguo)". *Contemporary China Studies (dangdai zhongguo yanjiu)* **81**(3). http://www.modernchinastudies.org/us/issues/past-issues/81-mcs-2003-issue-2 /1288-2012-01-06-09-16-39.html

Xu, J., P. Zhang, et al. (2003). "Land Usage and Changes in the Region of Lancangjiang River, Yunnan (*yunnan lancangjiang liuyu tudi liyong he fugai bianhua)". Yunnan Vegetation Studies (yunnan zhiwu yanjiu)* **25**(2): 145–154.

Yang, G. (2003a). "The Co-evolution of the Internet and Civil Society in China". *Asian Survey* **43**(3): 405–422.

—— (2003b). "The Internet and Civil Society in China: A Preliminary Assessment". *Journal of Contemporary China* **12**(36): 453–475.

—— (2003c). "Weaving a Green Web: The Internet and Environmental Activism in China". *China Environment Series* (6): 89–93.

—— (2005). "Environmental NGOs and Institutional Dynamics in China". *The China Quarterly*, **181**: 46–66.

Yang, G. and C. Calhoun (2007). "Media, Civil Society, and the Rise of a Green Public Sphere in China". *China Information*, **21**(2): 211–236.

Yee, W.-H., C. W.-H. Lo, et al. (2013). "Assessing Ecological Modernization in China: Stakeholder Demands and Corporate Environmental Management Practices in Guangdong Province". *The China Quarterly* **213**: 101–129.

Yi, R. (2007). Yunnan Snub-nosed Monkeys: Please Give Them Our Attention for Once (*dianjinsihou zheci qing guanzhu ta). Science Times*

Yu, J. and S. Guo (eds) (2012). *Civil Society and Governance in China*. London, Palgrave Macmillan.

Zeng, F. (2009). "The Role of News Media in Contemporary Environmental Movements in China: from China's Environmental Protection New Century Tour to the Xiamen PX Protest" (*dangdai zhongguo huanjing yundong zhong de meiti juese cong zhonghua huanbao shijixing dao xiamen PX). Modern Advertising (xiandai guanggao)* **171**: 36–41.

Zhang, L. (2007). "Socialist Modernisation between Modernity and Utopia (*xiandai xing yu wutuobang jiaozhi zhong de shehui zhuyi xiandaihua)". Theory and Modernisation (lilun yu xiandaihua)* 3: 9–12.

Zhang, L., A. P. J. Mol, et al. (2007). "The Interpretation of Ecological Modernisation in China". *Environmental Politics* **16**(4): 659–668.

Zhang, S. and J. Chen (2009). "A Study of Water Shortage in Northern China (*huabei diqu queshui fengxian yanjiu)". Journal of Natural Resources (ziran ziyuan xuebao)* **24**(7): 1192–1199.

Zhang, W. (2007). "Green News and the Rise of Environmental Journalists in China (*lvse xinwen yu zhongguo huanjing jizhe qun zhi jueqi)". Journalists (xinwen jizhe)* 5: 13–17.

Zhang, Z. A. (ed.) (2010). *Inside Stories of 30 Years' In-Depth Reporting in China (qianru shenhai shendu baodao 30 nian muhou guiji)*. Guangzhou, Nanfang Daily Press (*nanfang ribao chubanshe*).

Zhao, J. (2014). "The Financial Account of Print Media" (*zhimei de zhangben). The Caijing State Weekly (caijing guojia zhoukan)*. Beijing. 6 January 2014.

Zhao, J. and Q. Xia (1999). "China's Environmental Labeling Program". *Environmental Impact Assessment Review* **19**(5–6): 477–497.

Zhao, S. (2004). *Investigating China*. Beijing, China Fangzheng Publisher (*zhongguo fangzheng chubanshe*).

Zhao, Y. (1998). *Media, Market, and Democracy in China: Between the Party Line and the Bottom Line*. Champaign, IL, University of Illinois Press.
—— (2000). "Watchdogs on Party Leashes? Contexts and Implications of Investigative Journalism in Post-Deng China". *Journalism Studies* **1**(2): 577–597.
Zhong, L.-J. and A. P. J. Mol (2008). "Participatory Environmental Governance in China: Public Hearings on Urban Water Tariff Setting". *Journal of Environmental Management* **88**: 899–913.
Zhou, J. (2013). "The Future and Prospect of the Shanghai Press Group (*qianlun shanghai baoye jituan de fazhan qianjing*)". *Southern Media Studies* (*nanfang chuanmei yanjiu*) **45**: 49–54.
Zhu, L. and Y. Long (2012). "Prominence and Control of Environmental Injustice in China (*zhongguo huanjing zhengyi wenti tuxian yu tiaokong*)". *Nanjing University Journal: Philosophy, Humanity and Social Science* (*dajing daxue xuebao zhexue renwe kexue shehui kexue*) **1**: 48–54.

Index

The manufacturer's authorised representative in the EU is Springer
Nature Customer Service Centre GmbH, Europaplatz 3, 69115 Heidelberg,
Germany. If you have any concerns regarding our products, please
contact ProductSafety@springernature.com

Printed and bound by CPI Group (UK) Ltd, Croydon, CR0 4YY
23/04/2026
02095595-0006